Contents

Acknowledgments

I owe the idea for this book to Celia Roberts. Many thanks to her for encouraging me to develop a reluctant and ambivalent interest in post-dot-com Internet cultures into something constructive. I continue to benefit from talking with colleagues at Lancaster University. Particular thanks are due to John Urry for healthy scepticism concerning mobile technologies, and Lucy Suchman for her critical acuity on all matters technological. Many people in the Centre for Science Studies, and Cesagen (Centre for Economic and Social Aspects of Genomics; a UK Economic and Social Research Council funded research centre), have conversed long and often about underlying themes of the book, especially those concerning the affects and promises of contemporary technoscience. Warm thanks to Maureen McNeil, John Law, Ruth McNally, Richard Tutton and Brian Wynne. Friends and colleagues in Australia, particularly Andrew Murphie, Gerald Goggins and John Jacobs, have been supportive in the face of the various theoretical and practical anomalies of this project.

The encouragement and incisive suggestions of initially anonymous readers of the manuscript have shaped the final version in significant ways. Thanks to Anna Munster and Chris Kelty for positive feedback.

Finally, I would like to dedicate this book to Connor and Callum.

1   Introduction

Motion, to take a good example, is originally a turbid sensation, of which the native shape is perhaps best preserved in the phenomenon of vertigo. (James 1996a, 62)

Between 1999 and 2009, a "turbid" or disordered sensation of change was felt as wireless connections expanded and eroded the edges of the Internet and mobile telecommunications. Wireless connections in the making were unraveling networks as the dominant fabric of contemporary media. A vertiginous, chaotic movement zigzagged across devices (routers, smartphones, wireless memory cards, netbooks, wireless radios, logistics tags, etc.), cities, diagrams, people, databases, logos, standards, wars, crimes, towers, Pacific Islands, Guangzhou workshops, service agreements, toys, states, bicycles, "exotic places" such as Timbuktu, theme parks and chip foundries. This book is a set of experiments in connecting movement of wireless to the "native shape" of turbid sensations. It draws on philosophical techniques that are almost a century older. The radical empiricism associated with the pragmatist philosopher William James offers techniques for matching the disordered flows of wireless networks, meshes, patchworks, and connections with felt sensations.

Wireless networks have been most often set up, used, configured, and figured in everyday locales such as railway stations, trains, airports, public parks, cafés, schools, and, above all, houses as a way of connecting to the Internet and the many modalities of communication that it supports—the Web, email, streaming audio and video, news groups, data transfer, voice and video calling, instant messaging, rss feeds, blogs, and so on. Wireless connections have found myriad uses—streaming audiovisual materials throughout the home, tracking children in Legoland, monitoring the growth of grapevines or wildfires through wireless sensor networks, participating in multiplayer console games, or more ambitiously, replacing corporate-owned telecommunications infrastructure with community-owned communications networks (PPA 2003). Well-known remote locations have been selectively targeted by wireless networks. Between 2001 and 2004, wireless networks for computer communications appeared in settings ranging from Everest Base Camp and Nepalese yak farms (Waltner 2003), to rural villages in Cambodia or the "first wireless nation," Niue in the South Pacific (St. Clair 2003).

The Insignificance of Wi-Fi   *wireless fidelity*

We recognize wireless devices, locations, connections, services, and networks in many guises, but perhaps above all in the form of Wi-Fi®. Wi-Fi is a registered trademark of the industry confederation of equipment manufacturers and service providers called the Wi-Fi Alliance: "The Wi-Fi Alliance created the term Wi-Fi, which has come to represent a cultural phenomenon" (Wi-Fi Alliance 2003). The Taoist-influenced yin-yang design trademark appears on many wireless devices and at many wireless hotspots. It "certifies" that devices for local area networking over several hundred meters meet particular technical standards; "consistent use of these [Wi-Fi Alliance] marks is fundamental to the growth and recognition of Wi-Fi products and services" (Wi-Fi 2009). Although the yin of "Wi" and yang of "Fi" probably do not mean very much in terms of the cosmic balance of opposing powers, the design of the logo signals the functional importance of "i." The lowercase "i" (particularly prominent in the names of Apple Corporation's recent products) defines the space around "Wi-Fi." The logo style guide specifies that the clear space around the logo shall be equal to three times the width of "I" (see figure 1.1). That seems apt. Wireless networks very much concern the interval between people, or the space around "I."

Figure 1.1
The Wi-Fi brand (Wi-Fi 2009)

Wi-Fi is a trademark for a high-speed local area networking standard developed during the 1990s by the Institute of Electrical and Electronics Engineers, an international standards-making body (IEEE 1999, 2003). A networking standard dating from ten years ago is in some ways a very unpromising candidate for network and media theory. Ten years later, many Wi-Fi devices surround us in urban settings, especially in homes, apartments, offices, and increasingly outside in public places, but how do we have any sense of them? There are numerous gadgets as well as software designed to detect Wi-Fi connections in the vicinity.[1] I bought a "Wi-Fi watch" that carries what looks like the logo of the Wi-Fi Alliance (I Want One Of Those Ltd, 2008) on the promise that it would make visible wireless networks in the vicinity. Although the watch tells the time, it is relatively ineffective as a wireless detector. It often does not detect the presence of a wireless access point, even when I am convinced there must be one there. When it does detect one, it does not say much about it. It displays signal strength as a number between one and eight without any other information—the name of the network, whether access to it is open or encrypted, and so on. It says nothing about how many other wireless access points are in the vicinity. The disappointing Wi-Fi watch, despite being well reviewed on various gadget sites on the Web, displays to an annoying extent many of the limitations of wireless devices more generally. The promise of wireless access to the Internet sometimes shrivels to an impassive demand for connectivity. The banal limitations of the watch consign it to oblivion in a drawer somewhere. However, its limitations might apply to wireless devices more generally. A certain undoing of networks occurs here. Wireless devices promise expanded network connectivity, but this expansion is provisional, uneven, and patchy. People are not always aware of the existence of wireless connections, nor do they need to be. Someone buys a new digital camera or mobile phone, and uses it for several months without ever noticing that it is Wi-Fi-enabled. The camera can connect to a wireless access point and upload images to another computer or photosharing Web site such as flickr.com. However, there is no need to do that, and many people might even be unaware of the possibility. So, it is very possible that soon very little will be known about wireless networks as they sink into a banal media-technological background.

Despite their insignificance and blandness (or perhaps because of it), wireless networks effervesce on the edges of media change, activating and catalyzing experiential modifications. The vigorous proliferation of wireless devices and products evidences some kind of ferment. For instance, in 2009, the online store Amazon.com displayed almost 22,000 different products under the category "wireless." These range from a Cisco WLAN (Wireless Local Area Network) controller worth tens of thousands of dollars to a GZ Wireless FM transmitter for an iPod or a Sierra Wireless 595U EVDO USB Modem, both selling for the inexplicably low price of £0.01, presumably because they are tied to service plans or other contractual arrangements. Despite the abundance of Wi-Fi devices—five billion Wi-Fi chips by 2012, according to market analysts (ABIResearch 2007)—the bare fact of network connectivity seems dull and listless. What amid this technological spindrift is worth comment? Across the face of this avalanche of wireless commodities, various degrees of openness, reconfigurability, or plasticity can be found. In general, Wi-Fi devices exhibit a high degree of openness to modification, hybridization, reconfiguration, and the widest variety of instantiations (commercial, personal, portable, citywide, environmental, etc.). Wi-Fi connections, intermittent, unstable, and uneven as they often are, act as a kind of patch or infill at the edges and gaps in telecommunications and network infrastructures. Wi-Fi is seen as most likely to practically deliver on the promise of the "Internet of things," the idea that all electronic devices will connect to the Internet (Itu 2006). Although many competing wireless "solutions" to the problem of connecting devices to the Internet can be found (Bluetooth, Zigbee, femtocells, pico-cells, 3G, LTE, WiMax), Wi-Fi continues to grow in popularity, partly because it is relatively cheap, and partly because it is "out of control" (that is, it requires little centralized infrastructural management).

Sometimes a single wireless device crystallizes something of the undoing and intensification of networks. It can superimpose waves of media practice and network culture on each other, with sometimes surprising effectiveness. A Wi-Fi Internet radio refers at once to the history of broadcast radio, Internet media and contemporary wireless culture. Historically, the very term *wireless* dates from the late nineteenth century, yet it languishes on the periphery of media studies and media theory. As Timothy Campbell argues, apart from voice-centered radio, wireless media have largely been occult in cultural and media studies (Campbell 2006, x–xi), even though wireless devices precede broadcast radio by several decades.[2] Radio communication and information networks have an intimate historical association. AlohaNet, a packet radio system developed in 1970 to allow communication between computers at seven island campuses of the University of Hawaii, furnished a basic approach to network traffic handling and collision avoidance still used in many parts of the Internet, and in particular, in all local area networks such as Ethernet (Abbate 2000, 115–117). The configuring of the Internet as space of reroutable flows of information owes something to wireless here. Finally, from 1999 onward, a set of very rapid changes associated with wireless devices affected how people use the Internet, where the Internet can be found, what kinds of devices belong to the networks, and indeed, what the Internet is for.

The key claim of this book is that the contemporary proliferation of wireless devices and modes of network connection can best be screened against the backdrop of a broadly diverging and converging set of tendencies that I call "wirelessness." Wirelessness designates an experience trending toward entanglements with things, objects, gadgets, infrastructures, and services, and imbued with indistinct sensations and practices of network-associated change. Wirelessness affects how people arrive, depart, and inhabit places, how they relate to others, and indeed, how they embody change. In floating such an awkward term as *wirelessness*, I would invite readers to attend mostly to the suffix *ness*. *Ness* seems to me to do a better job than *wireless* of capturing the tendencies, fleeting nuances, and peripheral shades of often barely registered feeling that cannot be easily codified, symbolized, or quantified. As a suffix, *ness* also tends to convey something about a state, condition, or mode of existence (*light-ness*, *heaviness*, *weak-ness*, *happi-ness*, etc.). In this respect, the messy, fragile, and often ill-suited aspects of Wi-Fi are actually useful. Wi-Fi's limitations and surprising potentials highlight wirelessness as a composite experience animated by divergent processes, by relations that generate transitions and create expectations of more change to come. "More" includes the "less" of wirelessness: there will be less wires, less obstacles, less difficulty, less weight, and in general much more of less. The diaphanous fabric of wirelessness spans several strands of media-technological change. The structure of this experience is diffuse, multiple and hazy in outline.[3]

KEY CLAIM

Peripherals and Peripheries: Wirelessness and Networks

Accepting that the banality and inadequacies of Wi-Fi are worth thinking about, what would this mean practically? In writing this book, the Wi-Fi watch was not the only form of minimal connectivity I encountered. Another connection problem ran across the desk I often work at. It concerns a wireless keyboard and mouse, licensed by the U.S. Federal Communications Commission to operate at radio frequencies around 27 MHz (Federal Communications 2004). Although the wireless connection between the mouse and computer, or keyboard and computer, works well most of the time, sometimes it mysteriously fails. Wireless mice often share radio spectrum with cordless phones, remote control toys, microwave ovens, as well as other Bluetooth devices. For a long time, I thought the metal in-tray next to the computer monitor somehow acted as an electromagnetic Faraday shield for the wireless mouse receiver near the computer. I moved the in-tray around a lot until one day crawling under the desk, I found that the receiver's plugs into the back of the computer were slightly wobbly, and that by touching them lightly, the keyboard and mouse started working again. Sometimes I have to crawl under my desk to touch the plugs running into the back of the computer.

Is crawling around in the dust jiggling plugs a matter of wirelessness? For a start, it suggests that there is no pure experience of wirelessness, no subject whose interiority could be the foundation or anchor point for such an experience.[4] Feelings of wirelessness are site-specific and attach to a mass of things, images, projects, products, enterprises, plans, and politics concerning not only wireless networks but forms of urban, economic, work, institutional, and everyday life more generally. So the "subject" who experiences wirelessness is not very salient. Wirelessness is not a strongly personal or intimate zone of experience, at least in the usual senses of these terms. The layers, intensities, resistances, and vectors of a wireless subject have a somewhat ephemeral and nebulous character. In all the variations in tendency and direction it triggers, wirelessness puts detours and obstacles on the path toward the point where all differences converge and coalesce in pure, total networks.

Many, if not all, readers of this book will have wide-ranging, firsthand experience of wireless connections. They will have sat in airport terminals and hotel lobbies using a laptop or some other device (BlackBerry, iPhone, netbook, etc.) to do e-mail or browse the Web. In houses and apartments, they might have set up or found a wireless network to access. They might have heard of a café shutting their wireless networks off on weekends to discourage all-day occupation of tables or they might have complained to their friends about the exorbitant cost of using in-room hotel wireless networks in business hotels. Someone living in the countryside might have established a long-distance wireless connection allowing them to work from home. Who hasn't felt annoyance and frustration at the sudden, seemingly random difficulty of connecting to a wireless access point? Or perhaps they will have noticed access-point lights flickering rapidly in the middle of the night when no one in the house is using the Internet.

Many people will have seen the accounts, anecdotes, and reports of wireless network use. Wireless networks such as Wi-Fi have been heavily discussed in electronic, print, and online media, especially as different kinds of wireless connections coalesce in single devices. So much media phosphoresces around wireless network infrastructures, devices, and products that it is difficult to isolate wirelessness as such. The news media have often reported on crimes, accidents, and dilemmas of wireless networking (for example, debating the ethics and legality of piggybacking on a neighbor's open wireless network). Occasionally, media attention has concerned crimes, terror, fear, and insecurity associated with wireless networks (for instance, the use of unsecured wireless access points to send e-mails by bombers in India in 2008, the theft of millions of customer credit card details from a U.S. retailer's wireless networks in 2006, and so on). Much more often, attention to wireless networking has taken the form of reviews, recommendations, and advertising of wireless devices and services.

These manifestations of wirelessness could be multiplied indefinitely without ever coming to a core problem of wirelessness. What are we to make of the morass of contemporary experience of wireless connections, devices, and networks in use? It sprawls across a diffuse, spatial, politico-economic, and, shall we say, philosophical periphery. Peripheries and peripherals are not without importance. For instance, much commercial and noncommercial attention, effort, thinking, and indeed feeling in wirelessness relates to versions of the so-called last-mile problem, a problem that many telecommunications operators face today. While central infrastructures and network backbones can mostly be constructed speedily and expediently, the sheer number and variety of connections needed to hook every room, desk, village, chair, building, bag, pocket, pole, cabin, footpath, or other place to a telecommunications network often entail vast expense, upheavals, or complications. From the late 1990s on, wireless networks and wireless techniques of various kinds have been championed as the most economical solution to this problem, a problem that actually ranges in scale from hundreds of kilometers (as in development projects in remote locations) down to centimeters (as in personal devices carried in a bag or pocket). The last-mile problem might seem peripheral to the turmoil of change associated with network media and communication in general. It seems that it might only concern practical problems of access and connectivity rather than the more absorbing, immersive experiences of working, shopping, communicating, playing, and socializing online. The latter certainly draw the attention of most critical work on networks and digital media. However, this peripheral aspect of wirelessness and wireless devices as peripheral to networks is precisely the focus of the book. Everything here begins at a periphery, a periphery that, like my cordless mouse, is often vexed by weak connections, the residual weight of wires, and barely felt interference.

Wire: A Struggle against Dispersion

resistance

Wirelessness struggles against wires, and the extensive tying and knotting of wires called "networks." In this struggle, the question is: Which wires go where, or who wires what? In 2002, Nicholas Negroponte, founder of MIT's Media Lab and long-standing contributor to *Wired* magazine, announced in an article titled "Being Wireless" that "everything you assumed about telecommunications is about to change. Large wired and wireless telephone companies will be replaced by micro-operators, millions of which can be woven into a global fabric of broadband connectivity" (Negroponte 2002).

The global fabric Negroponte envisions will be made of free-associating Wi-Fi access points. In 2009, many aspects of this change are tangible. Take the Cradlepoint PHS300 Personal Wi-Fi Hotspot, an ideal piece of technology for the mobile "micro-operator": "The PHS300 Personal WiFi Hotspot is a true plug 'n' play solution that creates a powerful WiFi network almost anywhere. Connect all your WiFi enabled devices by simply plugging in your activated USB data modem and turning on the PHS300. It's that easy! No more searching for a hotspot, you are one!" (Cradlepoint 2009).

Micro-operators do not search for or connect to networks, they make networks for themselves. The PHS300 fits a niche in the contemporary wireless ecology since it links Wi-Fi networks to 3G/4G mobile broadband services provided by telephone companies. It bridges the computer industry and the telecommunications industry. The hitch here is that the devices requires an additional USB data modem—that is, "a 3rd party data modem and active data plan." Instead of the complete replacement imagined by Negroponte, the PHS300 represents the interface between different wireless networks in the same space. Rather than a "global fabric of broadband connectivity," wireless networks became one style of connection among many. There are so many ways to connect: Bluetooth, Wi-Fi (a, b, g, n), WiMax, GSM, LTE, EVDO, 3G, 4G, and so forth. The average home user may discover half a dozen "wireless networks" in the vicinity, their mobile phone may also be a Wi-Fi and Bluetooth device, and their laptop computer may have a mobile broadband USB data modem that connects it to mobile broadband networks operated by telephone companies.

Any struggle against wires faces the fact and figure of existing networks, particularly the Internet but also those of commercial telecommunications operators, broadband service providers, and institutional network administrators on which the Internet actually relies. Networks of different kinds overconnect a given location.[5] Hence, wirelessness in all its contemporary

guises encounters a plurally networked universe. Moreover, the trope or
figure of the network as expansive relationality affects wireless devices
and infrastructures at almost every level. Since the 1980s, network-oriented theorizing has pounced on every scale, order, and variety of phenomena ranging from the subatomic to the global, and wrapped them in
network form. The standard sociological definitions of a network from
sociologists such as Manuel Castells and Jan Van Dijk are terse yet indefinitely expansive: "a network is a set of interconnected nodes" (Castells
1996, 470). They can also be slightly more restrictive: "a collection of
links between elements of a unit" (Van Dijk 2006, 24). Any number of
more or less adorned variations stress that relations come before substance, identity, or essence.

No doubt it is hard to shun networks as "the new social morphology of
our societies" (Castells 1996, 469) and as one of "the dominant processes
and functions in our societies" (Castells 1996, 470). In the course of this
book, I accept that it is impossible to understand wirelessness apart from
networks. It is hard to argue against a hard-won focus on relations and
relationality associated with network analysis. Furthermore, wirelessness
comes into being within network cultures, the Internet, software, and the
electronics industries. However, after a decade of heavily network-centric
social, cultural, organizational, and mathematical network theory, there
are reasons to begin to approach networks a little more diffidently. While
it exhorts attention to relations, network theorizing can deanimate relations in favor of a purified form of networked stasis. Much network theorizing expects networks to have well-defined links and to afford unmitigated
flow between distinct nodes. While pure flow might sometimes occur when
a lot of aligning and linking work is done, very often it does not. Network
flows are actually quite difficult to manage and to theorize. Media theorists
Anna Munster and Geert Lovink (2005) put the very figure of the network
into question. "Theorising networks . . . must struggle with the abstraction
of dispersed elements—elements that cannot be captured into one image.
The very notion of a network is in conflict with the desire to gain an
overview."

The figure of the network struggles to contain an always-already
abstracted dispersion. Despite the extraordinary contemporary investment
in networks as the epitome of the contemporary real, they are, as Munster
and Lovink (2005) suggest, "unpredictable, often poor, harsh, and not
exactly 'rich' expressions of the social." If "struggle with the abstraction of
dispersed elements" runs through networks, wireless networks might make
the internal conflicted desire to network more visible, and in that sense,

save the network from itself. As the sociologist Andrew Barry (2001, 16) also argues, "The social world should not be imagined and acted upon as if it were a system of networks and flows, which can be grasped and managed as a whole. This is a typically modern political fantasy. The specificities and inconsistencies of the social demand careful attention."

Wireless networking could be seen as one venue in which the contemporary world is both imagined as a network flow that can be grasped and managed, and yet falls prey to constant inconsistencies and interruptions. Barry highlights that networks will always be "a part of, and yet not contained by, other collective arrangements or networks" (p. 18). Inconsistency and lack of containment very much typify wireless networks as they float in a foam of other media, settings, and environments, including the Internet and cities.

Recent work on networks such as Shaviro 2003, Galloway and Thacker 2007, Terranova 2004, and Chun 2006 alloy an awareness of the minimalist sheen of network formalism with sober attempts to identify edges and creases where the fabric of the network frays, and the crisp figure of network relationality crinkles. I take as a crucial point of departure Steven Shaviro's (2003, 249–250) conclusion that "what's missing [from life in the network society] is what is *more than information:* the qualitative dimension of experience or the continuum of analog space in between all those ones and zeroes. From a certain point of view, of course, this surplus is nothing at all. It is empty and insubstantial, almost by definition. . . . But this *nothing* is precisely the point. Because of this nothing, too much is never enough, and our desires are never satisfied."

In that case, what remains to be done with the figure of the network? Alexander Galloway and Eugene Thacker (2007) urge attention to protocol in network analysis and network politics. Their notion of "protocological control" enjoins detailed attention to, for instance, the "question of how discrete nodes (agencies) and their edges (actions) are identified and managed as such." From a different angle, Tiziana Terranova (2004, 90) suggests that the glitz and dazzle of high-profile websites such as MySpace. com or Facebook.com should be understood in relation to the practices and pattern of network labor that animate them: "It is the labour of the designers and programmers that shows through a successful web site and it is the spectacle of that labour changing its product that keeps the users coming back." Or turning to the "user," the slightly careworn figure who lingers around any discussion of contemporary digital media, Wendy Chun (2006, 249) draws our attention to the ways many accounts of networks dim our awareness of the uncomfortable fact that "all electronic

interactions undermine the control of users by constantly sending invol-
untary 'representations.'" The network user is a troubled figure, plunging
and ascending between freedom and control, sometimes configured as
windowless monad, and sometimes a locus that disrupt all closures.

Wirelessness offers a chance to pursue network gaps, frictions, and
overloads, and to develop less pervasive, less expansive or ubiquitous
figures of being-with each other. Although they are certainly not the most
luminous hotspot of practices or changes associated with media technological cultures, wireless devices and wireless connections might undo
from within valorized, and inescapable discursive figures of the network,
just as they intensify and extend it. The commonly used term *wireless
network* has an inherent tendency toward both self-erasure and multiplica-
tion. If a network has no wires, if it has few wires, then is it still a network
to the same extent? Does the substitution of radio signals for cables and
wireless dissolve the figure of the network? Dynamics at the level of
network protocol, production practices, and user subjectification play out
in multiple forms in wireless networks. In some respects, wirelessness could
almost be nothing, just the bland, pure, insatiable connectivity represented
by the many available wireless networks or the many overlaid network
connections. Wirelessness affords no strong ontological affirmation of
networks or reassertion of the primacy of the network as the essence of
the age. I see the morass of wirelessness rather as an opportunity to "give
the network back to the world."[6] If wirelessness does not augment the
network as a figure of "a single and yet multidimensional information
milieu—linked by the dynamics of information propagation and seg-
mented by diverse modes and channels of circulation" (Terranova 2004,
41), if wirelessness does not readily support a "philosophy for new media"
(Hansen 2004) (although wirelessness is certainly dazzled by effects of
"newness"), what does it offer?

"Less": Pragmatic Patches and Fields

Enormous as is the amount of disconnexion among things. (James, 1978)

In the United Kingdom and United States, the presence of radio-frequency
waves emitted by wireless networks in schools, daycare centers, and homes
is increasingly regarded as problematic (see WiFiinschools 2009). This
attention echoes long-standing uncertainties around radio waves and electrical fields associated with electric power networks and mobile phone
towers. Wi-Fi, it seems, makes some people sick (Hume 2006). An episode

of the BBC television prime-time current affairs show *Panorama* titled
"Wi-Fi Revolution" frames the problem as the "martini-style Internet,"
"fast-becoming unavoidable," "but there is a catch: radio-frequency radiation, an invisible smog. The question is, is it affecting our health?" (BBC
2007). Whether wireless signals substantially modulate the expression and
regulation of certain proteins in neurons, whether they damage reproduc-
tive or immune system function, is open to ongoing scientific debate
(Blank 2009). Regardless of how that debate is resolved, the court actions
in the United Kingdom and United States, as well as the BBC program,
indicate that people experience an increasing density of wireless devices
and signals as affecting their bodies and their health.

Wireless devices and infrastructures create zones or fields of equivocal
and indistinct spatial proximity. Many events in recent years have broadcast awareness of this equivocal proximity. Publicity about war chalking,
the short-lived practice of marking the presence of nearby wireless networks on pavements or walls (Hammersley 2002), was an early sign. War
chalking continues in many online wireless mapping projects, ranging
from industry-sponsored maps (such as the Wi-Fi Alliance's "Zone Finder"
(Wi-Fi Alliance 2003)) to war-driving or war-flying maps. War chalking
crosses boundaries between public and private. In the last five years, there
has much debate, somewhat inconsequential probably, about the ethics
and legality of accessing open wireless networks. High-profile cases have
occurred. The fifteen-month sentence of a teenager in Singapore for using
a neighbor's wireless network to play games (Chua Hian 2007), the theft
of 45 million customer records via the wireless networks connecting check-
outs to back-office databases at T.J. Maxx stores in the United States
(Espiner 2007), and the conviction of a U.S. man who parked outside the
local coffee shop every lunchtime and checked his e-mail using a wireless
network advertised as "free and open" (Leyden 2007) suggest that the
equivocal proximities generated by wireless devices lead to many kinds of
uncertainties: What properly constitutes a network when its edges and
nodes tend to blur into a patch or a field of connections?

The intimacy of wireless connections is embodied in remarkably abundant and literal ways—for instance, Wi-Fi monitored cardiac pacemakers
now being tested in the United States (Shapiro 2009). People attune themselves to signal availability and signal strength as they move around the
world. Subtle and sometimes gross alterations in everyday habits form
primary components of wirelessness as experience. Wireless signals are
experienced unevenly—as sickness, as frustration, as opportunity, as necessity, as hope. People have an inchoate sense of how the signal-processing

algorithms in their many devices are expanding and multiplying relations, continually propagating signals outward in crowded urban settings, overflowing existing infrastructures and environments and realigning senses of personhood at many junctures and on different scales. They are strangely composite or mixed experiences of indistinct spatialities. They trigger various attempts to channel, amplify, propagate, signify, represent, organize, and visualize relations—antennae are modified to exponentially increase the range of networks (Greene 2008), new software is coded to replace the preinstalled software on consumer wireless equipment (OpenWRT 2008), databases of wireless nodes appear on the Web (WeFi.com 2008), and Wi-Fi fractalizes into hundreds of minor and major devices, applications, and projects. Despite its mundane existence, or perhaps because of it, Wi-Fi displays constant contractions and dilations, and multiple instantiations, sometimes awash in broader shifts in mobility, sometimes stretched or frayed by contact with other infrastructures, media, and events. Its plurality and lack of coherence offers a way to tap into networks-in-formation.

Any attempt to make sense of the banal plasticity and abundant unfolding of wireless networks, I would argue, would benefit from a pragmatist approach, or at least, from that variant of pragmatism associated with William James called "radical empiricism." Pragmatism argues for no particular *Weltanschauung* or worldview. It has therefore often been understood as a method of evaluating ideas in terms of their practical usefulness: "Consider what effects that might conceivably have practical bearings you conceive the objects of your conception to have. Then, your conception of those effects is the whole of your conception of the object" wrote (Peirce 5.438, 1878/1905). For James (1996b, 263), "What really exists is not things made but things in the making." Recent work on James as well as on Charles Sanders Pierce and John Dewey in various parts of architectural and cultural theory, science studies, and contemporary political philosophy probes the undercurrents of this (see Lapoujade 2000, Ockman 2001, Massumi 2002, Grosz 2005, Ferguson 2007, and Debaise 2007). What opens out in James's thought from this starting point is an exceptionally vivid conceptualization of the processes of moving, making, changing, altering, and connecting of feelings, things, events, images, textures, ideas, and places. As Didier Debaise (2007, 8) writes, "We would certainly not deny the importance of the [pragmatic] method which joins together a redefinition of experience and a transformation of modes of knowledge, but it seems to us to be animated by a more profound intuition which gives it meaning [*sens*]."

What is this more profound intuition? It could be expressed in many different ways. First, it concerns the relation between thinking and things. One of the key traits of James's (1996a, 125) radical empiricism is a plural conception of things: "One and the same material object can figure in an indefinitely large number of different processes at once." Things themselves belong to diverse processes. Elizabeth Grosz (2005, 132) observes: "As the pragmatists understood, the thing is a question, provocation, incitement or enigma. The thing, matter already configured, generates invention, the assessment of means and ends, and thus enables practice."

Indeed, rather than reducing thought, concepts, or perceptions to instrumental ends, as if thoughts were just tools for living, pragmatic thought problematizes the mode of existence of things as distinct from thought. "Things neither commence nor finish, there is no entirely satisfying denouement," writes Henri Bergson (1934, 241) in his preface to a volume containing James's essay "On Pragmatism," which had been published in France in 1911. A feeling of incompletion or openness that James attributes to all things is very present in wireless networking. The configurations of matter and energy in wireless networks are complicated. Their overflow provokes, or incites, for instance, court cases against wireless networks in schools. Their potential for reconfiguration incites practices, reassessment of means and ends, and inventions. In their many variations, from 2000 onward, wireless devices and infrastructures augured a more down-to-earth, located, field-tested, and service-packaged form of the relatively abstract schemes of the network or the virtual so popular in the 1990s.

So, pragmatic thought is not just provoked by things. It affirms the practical inseparability of thinking and things. They are only separable in principle, never in experience. As James (1996a, 37) says, "Thoughts in the concrete are made of the same stuff as things are." (This implies that this book itself is no different in nature from wirelessness—it is another process diverging out of wirelessness, one among many.) Thinking amid things is vital to pragmatism. While it runs deep throughout pragmatist thought (Peirce, James, Dewey), it expresses itself most richly in James's "radical empiricism." Why should things and thinking be so entwined? Every experience, including any thinking we might do as analysts, results from interaction between life and its milieu. As David Lapoujade (2000, 193) writes, "Experience must therefore be understood in a very general sense: pure experience is the ensemble of all that which is related to something else without there necessarily being consciousness of this relation."

The lightly structured, apparently subjectless account of experience proposed by James seeks to account for a certain overflowing, excessive, or propagative aspect of experience that occurs in the absence of any pregiven form. It focuses closely on "change taking place," on the continuous reality-generating effects of change, and on the changing nature of change. As James writes, "'Change taking place' is a unique content of experience, one of those 'conjunctive' objects which radical empiricism seeks so earnestly to rehabilitate and preserve" (p. 161). While notions such as "conjunctive object" remain to be developed, the key point here is that experience itself is immersed in shifting settings or sites.

The cultural theorist Brian Massumi (2002) highlights this strand of James's thought in his account of the *transcontextual* aspects of experience. In reflecting on James's reformulation of an expanded field of experience that includes things and thinking, Massumi describes the streamlike aspects of experience: "We become conscious of a situation in its midst, already actively engaged in it. Our awareness is always of an already ongoing participation in an unfolding relation" (pp. 13, 230–231). Experience overflows the borders and boundaries that mark out the patched-in principal lived forms and functions of subjectivity-self, institution, identity and difference, object, image, and place. To become conscious of what it means to be engaged in a situation is to discover what makes one part of its fabric or tissue. This implies particular attention to edges: "Experience itself, taken at large, can grow by its edges. That one moment of it proliferates into the next by transitions which, whether conjunctive or disjunctive, continue the experiential tissue, cannot, I contend, be denied" (James 1996a, 42).

As we have already seen, peripheries are central in wirelessness. Wireless networks and devices are in some ways merely pragmatic continuations of network transitions. However, at every level, from concerns about radiation hazards to hopes for a complete transformation of global telecommunication, this practical extension of networks also provokes and incites, it participates in widening circles of relations, and it draws attention to proliferating and unfolding edges.

"ness": Inconspicuous Tendencies and Transitions

Take for instance the waves of change associated with Wi-Fi as it has moved through different versions in the last ten years. In each version—802.11a, 802.11b, 802.11g, and now 802.11n—wireless networks changed. There were changes in rate as the rate of information transfer increased (sometimes by large factors). There were variations in direction as community networks, municipal networks, and wireless development projects deployed different tactics, plans, and processes. Wi-Fi technologies spread across a range of consumer electronics and audiovisual media, as well as competing with mobile phones. The access points or wireless routers connected to the telephone or wired network, the network cards, and the antennae still looked more or less the same, or became less visible. Many more gadgets (phones, cameras, music players, televisions, photoframes, radios, medical instruments, etc.) became wireless. Often transitions between different Wi-Fi access points and networks are only distinguishable on the basis of small changes in feelings of connectivity, in variations of celerity, in the rather minute and fleeting flashes of network and signal-strength icons. On other occasions requests to authenticate or pay for connection entail larger "variations in direction." Processes of setting up equipment and connections changed slightly, especially in relation to encryption and security controls, but also on larger scales of network management where new mesh topologies embody forms of collaborative work. What would radical empiricism do with all this?

Although not a particularly scientific or even social scientific empiricism, the empiricism at stake in James's radical empiricism holds that knowledge can be patched into experience more fully if it attends to change, edges, and proliferation. Radical empiricism seeks to tread water in the unfurling of change long enough to become aware of "already ongoing participation" in it. Does radical empiricism simply reassert the primacy of subjective experience, and hence of the subject of experience? While the notion of experience runs very broadly in James's thought, James does not make the subject's experience the foundation of everything else in the way that, for instance, early twentieth-century phenomenological thought did (Edmund Husserl's life world is a totality of experience, a "phenomenon of being" (Husserl 1965, 19)). Experience does function as a platform in James, but a shifting platform for experimentation, not a solid foundation (for instance, of Husserl's "transcendental ego").

A rule of method guides the use of this platform: "To be radical, an empiricism must neither admit into its constructions any element that is not directly experienced, nor exclude from them any element that is directly experienced. For such a philosophy, *the relations that connect experiences must themselves be experienced relations, and any kind of relation experienced must be accounted as 'real' as anything else in the system*" (James 1996a, 42).

The maxim of radical empiricism is to admit only 'directly' experienced elements. Experience serves diverse materials or elements that can be used in patching together knowledge. The criterion of directness is very expansive, since it does not specify who or what experiences. Importantly, for our purposes, what James says of things—that they participate in diverse processes—has a direct corollary in experience in general: "Experience is a member of diverse processes that can be followed away from it along entirely different lines" (James 1996b, 12).

Like things, experience belongs to diverse processes. In particular, experience can be followed into things, perceptions, ideas, feelings, affects, narratives, memories, and signs as well as institutions, inventions, laws, and histories. Debaise (2005, 104) describes the result of the application of the rule of experience: "Everything is taken on the same plane: ideas, propositions, impression, things, individuals, societies. Experience is this diffuse, tangled ensemble of things, movements, becomings, of relations, without basic distinction, without founding principle."[7]

These processes might be commercial, political, personal, organization, military, governmental, leisure, educational, scientific, environmental, industrial, logistical, and so on, in nature. All of these figure in wireless networks at different points. Wireless networks enter into "a large number of different processes" at once. If we accept that experience and things are deeply coupled in the ways suggested by James, wirelessness appears as a composite or mosaic experience, a "member of diverse processes." Wirelessness as experience runs across boundaries between technology, business, politics, science, art, religion, everyday life, air, solid, and liquid. In the light of James's expanded notion of experience as expansion and divergence in full flow, we would need to ask: What diverse processes does wirelessness belong to? A radical empiricism that lived up to its promise would need to engage with experiences ranging from the infrastructural to the ephemera of mediatized perception and feeling. It would find itself moving across a patchwork of exhortatory hype, gleaming promise, highly technical gestures, and baffling or bland materialities.

Rather than being directed toward the endpoint of endless, seamless, ubiquitous connectivity of all media and the pervasiveness of the Internet, we might begin to attend to ways in which wirelessness alters how transitions between places occur. Wireless hotspots are set up in cafés, hotels, trains, aircraft, neighborhoods, parks, and homes. Sometimes they promise to generate revenue or make mobility easier for commuters or travelers. However, the question in all such cases is how such transitions introduce,

WLN = EXP.

transitions
b/t
places

as James puts it, "variations in rates and direction." The kind of awareness that might emerge from radical empiricism is distinguished by an experimental interest in tendencies. Radical empiricism does not offer much by way of an ontology, let alone a worldview of a subject. Rather it is a way of inhabiting transitions. James (1996a, 69) writes: "Our experience, *inter alia*, is of variations of rate and of direction, and lives in these transitions more than in the journey's end. The experiences of tendency are sufficient to act upon."

This sounds incredibly general, a flat truism that is hard to disagree with. Is James saying that we inhabit "transitions" more than ends in general, just like people often say, rightly perhaps, that the journey is more important than the destination? James offers more than an opinion on the value of experience. The feeling of continuous or discontinuous transition is, for him, what gives consistency to any experience, what allows it to flow. This feeling of change, transition, or in particular, tendency, is the fabric of any experience of acting or being acted up. Empiricism is radical to the extent that it manages to hold onto "the passing of one experience into another" (p. 50), and this means holding onto tendencies, tendencies that operate like differentials expressing rates and directions of change. The key challenge in adhering to the rule of direct experience is accessing these kernel transitions. They can be minimal, since they tend play a quasi-infrastructural role. The continuity of transition relies on, as James (1996b, 96) puts it, "the through-and-through union of adjacent minima of experience, of the confluence of every passing moment of concretely felt experience with its immediately next neighbors."

Tendencies have something profuse, overflowing, or excessive in them. James's emphasis on variations of rate and direction target apparently nonpractical, nonuseful, excessive, or irrelevant components that appear only as tendencies or potentials in experience. If experience comprises "variations of rate and direction," and if these variations are lived as the passing or transitioning of experience, what can be done with them? How can they be evaluated? How can they be known?[8] How can they be prolonged enough to register consciously?[9]

Conjunctive Relations: Movement and Stasis

In 2008, IEEE 802.16 WiMax (Worldwide Operability for Microwave Access) (IEEE 2004) could be said to embody the state of the art in wireless mobile communication infrastructure. WiMax differs from Wi-Fi in many ways (for instance, it mainly uses licensed spectrum rather than unlicensed spectrum). It markedly increases the range and speed of network connections, and has quickly been adopted as a way of building "last-mile" network infrastructures in places such as Greece (wimaxday 2008), Vietnam (Vietnam News 2007), Chile, Tanzania (Gardner 2007), Macedonia, Uganda, Bolivia, Turkey, and Nigeria (Patrick 2008). For instance, the "Holy Mountain" of Athos, a self-governing monastic state in northeastern Greece, is home to twenty Eastern Orthodox monasteries. The peninsula, accessible to male-only visitors by boat, is now blanketed with high-speed wireless WiMax coverage powered by solar cells and wind turbines. Mount Athos's monasteries embody a contemporary extreme of stasis and immobility. Yet, as Castells et al. (2007, 248) argue, in contemporary mobile communication, mobility is indexed to stable locations: "The key feature in the practice of mobile communication is connectivity rather than mobility. This is because, increasingly, mobile communication takes place from stable locations, such as the home, work, or school. . . . Mobile communication is better defined by its capacity for ubiquitous and permanent connectivity rather than by its potential mobility."

This statement is meant as an antidote to industry-driven mobility hype that often treats business and work travelers on the move through departure gates as the embodiment of mobile communication. Much more often, we could easily imagine, mobile communication entails finding a place where one can stop moving and sit—a café table, a park bench, a sofa, or some other form of seating. From this perspective, Mount Athos and its WiMax monasteries instance a relatively long-stay form of seating, not an exception to the trend toward mobile communication.

James would argue that even the most rigid stasis is deeply mobile. Experience constantly passes through many different states, ranging from the impersonal to the personal, from the singular to the general. On any scale we imagine, there is no pure flow or pure sensation of transition. Many transitions occur between scales. And every transition is shot through with temporary termini, with snags, resistances, circularities, and repetitions. James (1996a, 4) often speaks of "pure experience" as his main methodological postulate: "My thesis is that if we start with the supposition that there is only one primal stuff or material in the world, a stuff of which everything is composed, and if we call that stuff 'pure experience,' then knowing can easily be explained as a particular sort of relation towards one another into which portions of pure experience may enter." However, this primal material is never experienced as such, since it always partitioned and marked by practices of making, doing, marking and knowing. As David Lapoujade (2000, 193) comments, "Pure . . . points to an intermediary reality outside of any matter/form relationship." Nonetheless, even if we accept that pure experience in radical empiricism is a kind of methodological postulate, or even "a plane of immanence" as Gilles Deleuze and Felix Guattari (1994, 46) call it, how would such material involve movement, transitions, or variations?

Much hinges here on attending to conjunctive relations. Any radical empiricist attempt to know transition more intimately, to accompany it, or to follow it to its limits, pivots on conjunctive relations. What are they? "With, near, next, like, from, towards, against, because, for, through, my— these words designate types of conjunctive relation arranged in roughly ascending order of intimacy and inclusiveness" (James 1996a, 45).

How then do conjunctive relations help us rethink wireless technologies against their ingrained tendencies toward utility or means? Crucially, conjunctive relations allow transition to occur. Without a *with* or *near* or *toward* or *through*, there can be no tendency and hence no transition. These felt transitions are crucial to "nature" or "whatness" in radical empiricism. All experience unfurls in diverse conjunctive relations. No "knowing" or "doing" can happen without a patchwork of such relations since they are the fabric of transition in general. Felt transitions are neither spontaneous, random, nor completely ordered. The patterns, paths, and trajectories of this passing must include variations in rate and direction.

We have already seen that James treats experience as a "member of diverse processes." As a composite, it is replete with variations in rate and direction. We have seen too that thinking and things are two sides of the same coin for radical empiricism. Now we are in a position to understand why this goes beyond a simple pantheistic assertion of the composite texture of experience, and the inmixing of things and experience. Experience is composite, diverse, inclusive of things and thinking, because it owes more to transitions and tendencies than to endpoints. All of this rests on a very specific treatment of the materiality of transition. "Radical empiricism," James writes, "takes conjunctive relations at their face value, holding them to be as real as the terms united by them" (p. 107). Conjunctive relations embody the differentials of proximity, distance, intersectionality, divergence, and delay. Coming into language, conjunctive relations are voiced by particles such as *with, between, in, before, far,* and *so forth.*

These relations are encountered incessantly. Life is lived far more in these relations than in the disjunctive relations associated with things or entities. As James (1996a, 237) writes, "While we live in such conjunctions our state is one of *transition* in the most literal sense. We are expectant of

a 'more' to come, and before the more *has* come, the transition, nevertheless, is directed *towards* it."

As a form of knowing, radical empiricism pivots on analysis and untangling of conjunctions. We should recognize that both thinking and things
work with and process via conjunctive relations. Wherever we look, we
find conjunctions, that aspect of experience that triggers expectations of
more to come, in movement. For instance, by sorting and reordering conjunctive relations of "before" and "after" in elements of a data stream,
signal-processing techniques handle the "severe channel conditions"
found in crowded cities. Or, by sorting and reordering relations of "inside"
and "outside," wireless infrastructures affect sensations of the presence of
others.

In its insistent attention to conjunctive relations, radical empiricism is
intimately stitched into experience, into impersonal, preindividual, and
intimate, subjectified dimensions of experience. In much of the following
discussion of conjunctive relations, we will see this conjunctive underside
of experience overflowing psychological or perceptual engagements. It
pervades organizations, institutions, transactions, apparatus, and infrastructures, and it propels a gamut of improvisations, temporary fixes, and
modifications associated with wireless communication and networks. It
brings different scales into surprising contact. For instance, many people
might say that they have no interest in, let alone experience of, the algorithmic signal-processing techniques implemented in wireless networks
such as Bluetooth, Wi-Fi, or 3G cell phones. Despite that, their sensations
of connection, their awareness of service availability, and their sometimes
conscious preoccupation with connecting their wireless devices via service
agreements or other devices all derive from the handling of conjunctive
relations in data streams implemented in wireless signal-processing chips.
No doubt, the paths along which sensations of transition move are highly
complex, and cross multiple scales. The passing of experience can take very
circuitous routes. Substantive and disjunctive relations backfill the vectors
of transition. There are many circuitous conjunctive relations present in
wirelessness.

Being-with as Thinking of Conjunction

Conjunctive relations present different degrees and kinds of inclusiveness
and intimacy. However, in the transcontextual situation of wirelessness,
ordering these degrees is problematic. There are two aspects to this problem.
The first is the network. The infrastructural and media conditions under

which James developed radical empiricism were much less extensively networked than today. A key question here is: What happens to radical empiricism under network media conditions? The second aspect is political economy. It is hard to find any analysis of work, capital, economy, commodity, or markets in James.[10] How can we develop a network media aware, politico-economic edge in the free flow of radical empiricism? The scope to do this already exists in radical empiricism. As James (1996a, 94) puts it: "Experience now flows as if shot through with adjectives and nouns and prepositions and conjunctions. Its purity is only a relative term, meaning the proportional amount of unverbalized sensation which it still embodies."

In certain respects, we could see wireless technologies as attempts to extract, organize, channel, and protract the ambient "unverbalized sensation" of conjunctive relations. The conjunctive relations it harnesses (with, near, beside), are surrounded by "adjectives and nouns and propositions" that attach flows of transitions to personal attributes ("your," "freedom," "together"). What shoots through the flow of experience—"experience now flows as if shot through with . . ."—complicates that flow considerably. Although radical empiricism is a general philosophical technique for experimentation with conceptual constructs, it can be reconfigured and oriented to the proliferating, intermediate-level, and multiscale relations present in wirelessness. This means attending to the play of "intermediate factors." In their overview of ethnographic work on contemporary capital, economies, technology, and work, the anthropologists Melissa Fisher and Greg Downey (2006, 24) observe: "One thing that emerges from an ethnographic mode of theorizing is a greater awareness of intermediate-level factors—institutions, legal standards, technical limitations, social alliances, supply chains of particular commodities, subcultural identities, communities of shared skills—in the adaptation of technology to human use."

Under network conditions, the orderings of intimacy and inclusiveness that characterize the flow of experience can become unstable. When networks launch topological reorderings of place, "with" or "near" can become confused with "my" or "for." And it is precisely this instability in orderings of intimacy that invites many different attempts to inject verbal, visual, commercial, and legal orderings into conjunctive flows.

A radical empiricism of wireless networks would need, from this perspective, to take an interest in service plans, node databases, consumer electronics product reviews, or public-private partnerships for wireless network development. It might need to think about how antennae and algorithms work to permit people to walk around a city. In moving through any of these intermediate levels, the technique of radical empiricism loops through "relations that connect," or conjunctive relations. As James (1996a, 42) says, "The relations that connect experiences must themselves be experienced relations, and any kind of relation experienced must be accounted as 'real' as anything else in the system."

The question then become how to loosen up the tight bounds of what counts as "our" experience of wireless enough to bring different scales of relation into "the system." New, surprising, or affectable senses emerge in various ways, but mostly in making something. As James (2004) says, "We patch and tinker more than we renew." In relation to wirelessness, the praxis of making ranges across literature, art, politics, publics, science, design, economics, media, and the military. There is contemporary wireless fiction (Doctorow 2005; Ryman 2004), wireless art (Savicic 2008; Kwastek et al. 2004), wireless politics (BBC 2003; Bureau d'Etudes 2007), and wireless design (Dunne 2005). While making things wireless differs widely across these different domains, radical empiricism can recruit praxes of making to unearth the limits of experience. Practices of making explore different scales and across scales. At each scale, and between scales, certain verbalizations and partitions are inscribed. They inject aspects, interfaces, systems, maps, sets, groups, plans, images, transactions, services, and infrastructures into the conjunctive situation. From the standpoint of radical empiricism (although it is not a standpoint, it is a treatment), these injections always tend to relaunch conjunctive relations outward, in more or less rigid, static forms. In some cases, making things wireless effects a capture of conjunctive relations in commodity form. In others, it expels them.[11]

While the following chapters will pursue the multiscale mode of existence of wirelessness in more detail, there is a broad-ranging exteriorization of conjunctive relations that can be expressed at a higher-scale ontological level, and in a way that strongly links capital, network, and connection. There is a philosophical reading of the historical circumstances of a particular conjunctive relation, "with," that I find helpful in understanding what is at stake in the exteriorization of conjunctive relations. In a series of writings over the last decade, the philosopher Jean-Luc Nancy has explored the connections between Being and capital (Nancy 2000, 2007). His analysis pivots on the problem of how to think of capital as a potent, corrosive determination of Being today.[12] He regards capital as a historical process of stripping away any foundation in meaning, substance, or subjectivity back to the conjunctive "with": "Capital exposes the general alienation of the proper—which is the generalized disappropriation, or the appropriation of misery in every sense of the word—*and* it exposes the stripping bare of the *with* as a mark of Being, or as a mark of meaning" (p. 64).

Being without foundation, capital is unbalanced. At the same time, however, capital exposes a vertiginous absence of ontological foundations. Capital brings, for Nancy, an opportunity to become sensible to the barest of all conjunctive relations, what he terms "Being-with." This relation cannot be appropriated, expropriated, alienated, stripped away, or expelled since it has no meaning or interiority as such. It is already alienated as such. Nothing that exists today, in any context—social, religious, economic, technological, scientific, aesthetic, psychological, cultural—can alienate its own alienability, an alienability that stems from the conjunctive relation of "with." As Nancy writes, "If Being is Being-with, then it is, in its being-with, the 'with' that constitutes Being; the with is not simply an addition" (p. 30).

In the contemporary setting, communication embodies Being-with: "In fact, [what is exposed] is the bare and 'content'-less web of 'communication.' One could say it is the bare web of the *com-* (of the *telecom-*, said with an acknowledgment of its independence)—that is, it is *our* web or 'us' as web or network, an *us* that is reticulated and spread out, with its extension for an essence and its spacing for a structure" (p. 8).

While I am wary of investing too much in Nancy's emphatic account of "us," the "*com-*" (or with) expresses in a distilled form, it seems to me, an exterior limit for conjunctive relations in wireless networks. Handled carefully, Nancy's work (along with other related work), in short, allows a specific retrofitting of radical empiricism; it builds out an edge that engages with the expansive operations of capital and its spaced-out productions of value. Equipping radical empiricism with Nancy's conception of the Being-with helps pay attention to the ways in the conjunctive relations are exteriorized.[13]

Determinate Ambulation and Chapter Order

Devices, infrastructures, crimes, court cases, service plans, security, development projects, international standards, spectrum licensing arrangements, wires, antennae, settings, configurations, websites, logos, login screens, and many, many places: as I have said, none of this seems stunningly promising or significant material with which to develop a sense of contemporary network cultures. However, in its burgeoning irreducibility to the figure of the network, in its tendencies to redraw edges and peripheries, to ramify and to percolate, in its transitions and variations, in its discomfiting of locations, boundaries, and partitions, and in its many corporeal, legal, physical, architectural overflows, wirelessness bundles tendencies that lend themselves particularly strongly to a radical empiricist analysis of the present. These tendencies take networks into the world along many paths, and at times, render the patchiness and bareness of networks in the world more palpable. Each of the following chapters tracks a single tendency of wirelessness, beginning with attempts to create wireless cities, and ending with quasi-global belief structures concerning the expansion of wireless worlds. In moving from city (chapter 2), through signals (chapter 3), devices (chapter 4), maps (chapter 5), products (chapter 6), world (chapter 7), to belief in wirelessness (chapter 8), the chapters assay conjunctive relations as they undergo adjectival qualification, partitioning, and externalization.[14]

Movement in the city is a key motif for radical empiricism and pragmatism more generally. "Cognition," write James (1975, 81), "whenever we take it concretely means 'determinate ambulation.'" Cities generate particularly intense and varied forms of concrete movement, and strong beliefs and imaginings of controlled movement. Much arriving and departing occurs in them, and on different scales. The movement of people and things in cities has occasioned much analysis in urban studies of mobility, and also much more widely in social and cultural theory. Chapter 2 seeks to locate the most significant figures or forms of movement associated with wireless networks in cities such as London and Taipei. The chapter focuses in particular on the organizing notion of the "wireless city" in London between 2002 and 2007. Although the idea of a wireless city takes various forms, it always seeks to reorganize patterns of urban movement. Rather than trying to identify or convey urban experience directly (as much urban sociology has done) or to converge wireless network with ideas of cities as flow, I situate the idea of the "wireless city" in relation to moments of transition irreducible to the figure of the network. These transitions are strongly felt, and therefore offer multiple paths for a radical empiricist approach. The idea of the wireless city is an attempt to organize and control wirelessness in cities. Over fairly brief spans of time, the idea of the wireless city is affected by substitutions that erode, replace, overflow, and redefine its edges. The pattern of substitutions and site-specific alterations in the idea of the wireless city is an integral component of wirelessness. If ideas such as the wireless city are lived, they are lived to the extent that their apparent fixity allows and even encourages substitutions of various kinds.

Chapter 3 turns to the crucial technical terrain of wireless chipsets and wireless digital signal processing (DSP). Why bother with the sometimes bafflingly complicated techniques and architectures of DSP? Urban life produces problems of noise and interference that lie at the very heart of wirelessness. The complications of DSP are one strong index of the conjunctive patching of wireless networks into urban settings. The history of wireless cultures could be written as a history of how to inhabit urban (and military) electromagnetic environments. As contemporary wireless techniques began to gel around digital signal processing in the 1990s, they confronted the problem of how many signals could pass through the same space at once without becoming mixed up or clashing with each other. Effectively, a wireless signal must tolerate the presence of many others, in the same way that the inhabitants of any city must learn to live amid many strangers. How can the presence of many others be tolerated in communication? How many others can be present before communication is baffled? How can changes in position, in disposition, in concentration and dispersions, in rate and direction associated with movement in cities be accommodated? The digital signal-processing techniques that underlie wireless communication in almost every contemporary form (as well as many other audiovisual media such as video, photography, and sound) display daunting mathematical sophistication and intricate computational processing. Embedded in wireless chips, digital signal processing keeps afloat bodily sensations of movement, lack of constraint, absence of tether and encumbrance. While there is no unmediated access to signal processing, I argue that certain aspects of wireless signal processing tinge contemporary perception with specific conjunctions of movement and of direction. A conjunctive envelope grows around movement and perception. Even the most windowless wireless monad, fully equipped with a connectivity decoupled and unplugged from any particular place, is hedged in by intricately patterned consensus.

Chapter 4 theorizes work done on devices such as routers, antennae, and other wireless components. As something to be worked on, as more or less tangible things, wireless devices occupy an unstable position in the ecologies of infrastructure, architecture media, and consumption. In some respects they belong to network infrastructures. They can be configured as extensions of network culture. In other respects, they act as highly wrought furnishings for monadic bubbles of communication. They fungibly embody cascades of change in the design, making, organization, ownership, and maintenance of existing communication infrastructures and network media. At the interface between network cultures and communication infrastructure, the chapter examines how devices undergo processes of variation and modification that make these instabilities sensible. This work of variation and modification circulates around standards such as IEEE802.11, but it also brings into play open-source software such as Linux. The standards embodied in wireless devices are not monolithically coherent entities. The standards themselves contain instabilities, gaps, and incompatibilities that trigger new variations. Existing notions of infrastructure as the background of experience cannot accommodate the dynamisms of physical, electronic, and software modifications of consumer wireless equipment. From the radical empiricist standpoint, these modifications inject site-specific mutations into the texture of experience. The fringes of perception tinged by wireless signal processing unfurl here via work done on opening, reconfiguring, and reconnecting devices.

Chapter 5 focuses on the management of wireless networks. It tracks attempts to concretely act wirelessly. The chapter focuses on practices of network building, network mapping, and connecting to networks as they appear at two edges of wirelessness: online wireless node databases and antenna modifications. Node databases have been at the center of community, free, alternative, commercial, and public-private wireless networking projects. They mostly record the geographic location of wireless antenna. Putting antennae in different places—higher, lower, inside, outside, closer, and so on—has been a critical step on the path to wireless networks (not just for Wi-Fi, but for all wireless technologies, ranging from mobile phones to RFID chips). In what sense does making a network, mapping a network, or connecting to a network involve node databases or antennae? How do we make sense of the different aspects of wireless networks expressed in node databases and antennae? Unlike much network theory, I have sought to circumvent the figure of the network in analyzing wireless networks. For reasons discussed above, this chapter remains somewhat skeptical and wary of networks as analytical devices. Instead it asks, when and how are collections of relations concatenated into networks? Are there actual wireless networks, or only tendencies to network and practices of networking? In the latter case, we can expect networks to take very different forms depending on the ways collections of relations are handled. In the course of analyzing some different node databases and their different network forms, the notion of practice or action itself comes into question. From a radical empiricist standpoint, everything has to be viewed from the standpoint of practice. However, as discussed above, there are antipragmatic practices, practices that patrol and cordon off the buried potentials in any situation for displacement, divergence, and decoupling of contexts.

Nearly every instance of wirelessness has product-related dimensions. Chapter 6 treats wirelessness in relation to products, and the claims, promises, skepticism, awareness, and incredulity associated with wireless products. Wirelessness is pitched in the form of products, as goods and services. What would a radical empiricism do with wireless products? Where is there room for experiment, for construction, for testing of ideas or claims in relation to a tumult of wireless products? The key claim of the chapter pivots on James's claim that the difference between inner and outer aspects of experience depends on a sorting process. "Things" appear as inner or outer depending on how they act on their neighbors. Using techniques of listing and sorting of wireless products, the chapter analyzes how neighbors and neighborhoods of relations take shape. Each product gesticulates a set of promises. Promises—of connectivity, of ease of use, of freedom from constraint, of speed, of pleasurable sensations of touch, seeing, and hearing, and so on—bind the product. Strongly antipragmatist tendencies disguise the potential of products to germinate unexpected change. Yet the presence of many different promises and products in a neighborhood of relations tends to trigger uncontrollable jumps. These jumps blur and erase boundaries, and entangle products, services, and wireless forms of subjectivity. Importantly, maintaining equivocal differences between inner and outer, between pure promise and actual reality, serves the purpose of generating ongoing transitions.

Chapter 7 follows a strand of wirelessness associated with normative notions of "world," "globe," and development. The chapter discusses and contrasts a number of wireless development projects on different scales. Often the most advanced or adventurous wireless platforms are tested in Africa, South America, or South Asia. Since 1999, wireless networking projects have rapidly multiplied in many places and on many scales. These public, private, and public-private partnerships projects sometimes avow global ambitions (for example, "to connect the next billion people"). The arguments of this chapter are framed by existing critiques of technological development, and by a desire to understand the sheer number and variety of globally oriented wireless development projects from the perspective of a spectator. I regard these projects as made to be seen. The question then is how to situation their performance pragmatically. In what ways do wireless networks for global "others" vouch for or validate "our" connectivity? In what ways do wireless networks verify the plural and uncontainable overflows of any form or figure of world or globe? The coexistence of very different scales of projects creates analytical difficulties that a radical empiricist approach can usefully address.

The final chapter stands back from the different approaches taken in the preceding six chapters. It reframes these approaches in terms of belief in wirelessness. How does a radical empiricist account of the contemporary moment as framed by networks, connections, services, and devices differ from other work on the present, on recent pasts, and on near futures? Here James's account of belief (in conjunction with Henri Bergson's concept of duration) plays a distinctive role. Framing belief in terms of liveness, momentousness, and force allows us to pose the question of what possible relation we can have to change. The discussion contrasts James's radical empiricism with its concentration on conjunctive relations and the overflow of experience with Bergson's method of intuition and its concern with "true" and "false" problems. This contrast differentiates a pragmatist sensibility attuned to the vertigo and turbidity of change. In particular, this chapter offers an alternative reading of the network condition of experience of nothing.

In sum, the pleonasm *wirelessness* designates (1) a sensibility attuned to a proliferating ethos of gadgets, services, opportunities, and enterprises that transmit and receive information via radio waves using Internet-style network protocols; (2) a strong tendency to make network connections in many different places and times using such devices, products, and services; and (3) a more or less heightened awareness of ongoing change and movement associated with networks, infrastructures, location, and information. Wirelessness in contemporary media and cities links directly to the core of James's argument concerning the flow of experience, and this seems a good place to start following the wireless lead.

2   Substitutions: Directions and Termini in Wireless Cities

In such a world transitions and arrivals (or terminations) are the only events that happen, though they happen by so many sorts of path. The only function that one experience can perform is to lead into another experience; and the only fulfilment we can speak of is the reaching of a certain experienced end. When one experience leads to (or can lead to) the same end as another, they agree in function. (James 1996a, 63)

Cities such as London, Montreal, Mumbai, San Francisco, Sydney, and Taipei offer a crucial set of leads in thinking about contemporary wireless experience in terms of transitions, arrivals, and terminations. Intimately coupled to the architectures, temporalities, tempo, embodiments and socialities of cities, wireless networks bring changes associated with neoliberal, mobile, global, network capitalism as well as a sense of potentials and tendencies irreducible to it. Wirelessness might be well be conceived as the "sorts of path" that appears when transitions and arrivals happen more often. In the name of connecting to the Internet anywhere and anytime, cities have been rapidly equipped as wireless platforms and infrastructures, with some spectacularly divergent consequences. Since the year 2000, many cities around the world have launched themselves as "wireless cities," and sometimes no less ambitiously, as "muni-wireless" cities. The very idea of a "wireless city," or the more North American "muni-wireless" city, only makes sense in the wake of the growth of the Internet in the 1990s. However, wireless connections in the wireless city quickly turned out to be irreducible to the Internet. If the idea of the wireless city animated the growth of wireless connections in cities around the world, the very idea of a wireless city also provides an opportunity to see what a pragmatic approach to contemporary wirelessness means in situ.[1] A series of substitutions occurred as the idea of a wireless city evolved in the years 2003–2008. *Substitutions* are, as we will see, vital to radical empiricism's understanding of the life of ideas.

Ideas as Pathways and Terminals

Because radical empiricism treats experience as a kind of ambulation that concatenates multiple overlapping relations, it is an implicitly urban-ready philosophical technique. In radical empiricism, and pragmatism more generally, ideas have an immanent function in concatenating and short-cutting transitions. They are not regarded as different from or above the rest of experience or life. Indeed James often refers to ideas as "conceptual experiences." Conceptual experiences substitute rapid transitions for slow ones:

> As a matter of fact, and in a general way, the paths that run through conceptual experiences, that is, through "thoughts" or "ideas" that "know" the things in which they terminate, are highly advantageous paths to follow. Not only do they yield inconceivably rapid transitions; but, owing to the "universal" character which they frequently possess, and to their capacity for association with each other in great systems, they outstrip the tardy consecutions of the things in themselves, and sweep us on towards our ultimate termini in a far more labor-saving way than following of trains of sensible perception ever could. (James 1996a, 64)

Ideas, then, present "paths to follow"; they offer determinate trajectories. While James develops this understanding of "ideas" as rapid transitions in opposition to transcendentalist accounts of ideas, radical empiricism more generally connotes an awareness of tendencies and potentials within experience that lead to experiments and variations in movement. Ideas particularly afford experimentation because they have "a capacity for association with each other in great systems." This capacity for association supports their specific function in "knowing" things. Ideas "know" things to the extent that they "terminate" in them. Arrival or termination requires a series of "conjunctive experiences of sameness" that continue and corroborate transitions: "In this continuing and corroborating, taken in no transcendental sense, but denoting definitely felt transitions, *lies all that knowing of a percept by an idea can possibly contain or signify.* Wherever such transitions are felt, the first experience *knows* the last one" (James 1996a, 56). The process of continuing and corroborating that James stresses here is concrete. It involves making relations over time and constructing intermediaries that allow experience to tend in particular directions.

Something broadly similar might be said of wireless connections. From the end of last century on, they came to be seen as offering highly advantageous paths to follow in expanding and ramifying the Internet into lives and places. They seemed to have the potential to associate with each in "great systems" such as the Internet, and they promised to sweep people and things to their "ultimate termini" faster than other trains of perception or conduct. (Although the sense of the term *terminal* as "computer terminal" is rather dated, in wirelessness, it seemed, "terminals"—handheld devices, laptop computers, Wi-Fi/cellular phones, wireless game consoles, real-time location tracking tags, sensor nodes, and so on—would be everywhere in the city. Conversely, wherever terminals or termini in any sense of the term were to be found, so would the experience of wirelessness.) It was hardly surprising that the city itself should then become wireless. In late 2004, Taipei announced it would transition to a wireless city because it had installed a citywide wireless information network grid (see "know, don't" 2004) called "WiFly." In announcing Taipei's wireless mesh, supposedly at that time the world's largest Wi-Fi agglomeration, Mayor Ma Ying-Jiu laid down two fairly simple premises: "Taiwan is the best computer-hardware manufacturer in the world and more than 90 percent of wireless Internet ports are made here. Plus, because of the government's support of such a plan, Taipei is becoming a wireless city" (Mo 2005). For Taiwan's state planners, producing wireless hardware is economically important. Taipei needs to have the biggest wireless network in order to demonstrate its capacity to produce wireless hardware. Hence, with government support, Taipei's "WiFly" mesh signifies the excellence of Taiwan's electronics and IT hardware manufacture. Taipei, with 4,000 access points, was often cited as the leading Wi-Fi city in the world (Kim 2007). But how is a "wireless city" actually inhabited, especially in light of such an indirect and weak association between microelectronics production capacity, state power, and city life? Several years later the practical significance of the city's wireless network was in doubt. After two years in full operation, Taipei's "WiFly" network was used by only 30,000 subscribers in a city population of 2.6 million (Kim 2007). It seems that less than 1 percent of the population had begun to inhabit the wireless city. No significant mobilization of Taipei's inhabitants through the wireless mesh had occurred. The modest subscription costs are said to have been an obstacle (Belson 2006). More importantly, an ecosystem of free Wi-Fi hotspots in cafés (approximately 1,300 according to jiwire.com) throughout the city provided a reliable substitute for the citywide mesh. Even if these hotspots lacked the technically sophisticated connectivity of the WiFly mesh, even if they were not integrated into single citywide networks, their locations and availability made them more immediately present in the lives of Taipei's inhabitants.

Taipei's trajectory toward wirelessness on a citywide scale is not unique. Virtually the same vision of wirelessly networked cities appeared in many cities around the world in the years 2004–2007: Philadelphia, San Francisco, London, Austin, Berlin, Seoul, and so on. Taipei's wireless grid proved to be one in a series of wireless grids that appeared in Asia, North America, Europe, and parts of the Middle East and South America, growing strongly in cities such as Philadelphia, Bangalore, San Francisco, Austin, Glasgow, London, and lately, Shenzhen, Beijing, and Jakarta. These wireless grids, also called "clouds" or "meshes," were and are sometimes municipal government projects (Hellweg 2005), and sometimes the products of telecommunications enterprises such as AT&T in the United States or BT in the United Kingdom (Tropos Networks 2006). Very often, they relied on unstable partnerships and contractual service agreements between city governments and wireless Internet service providers. Like Taipei's, many of these citywide networks struggled for commercial viability, even as the popularity of wireless network access in cafés and other semipublic and public places grew. Some high-profile muni-wireless projects, such as San Francisco's Google-Earthlink Alliance (Kim 2007; Mills 2005) and Philadelphia's "Wireless Philadelphia" (Wireless Philadelphia Executive 2006) foundered even as Wi-Fi equipped people, devices, and places burgeoned throughout cities. Perhaps more importantly, in the same years, city wireless networks constructed by alliances of community groups, political activists, artists, and technologists flourished in cities such as Berlin, Montreal, New York, and London (Forlano 2008). How would we analyze the relative "failures" of the "wireless city" from the perspective of a radical empiricist account of wirelessness? Is a citywide wireless grid simply too ambitiously homogenizing for the low-power, popular practices of wireless network access? Attempts to remake cities in the form of grids are not new. Grids recur in the history of urban forms (Kostof 1992). At various levels, the idea of a grid of streets is quite familiar as a way of visualizing, organizing, and controlling movement in cities.[2]

The trajectory of any particular city toward wirelessness is difficult to attribute to a single actor such as the state, business, or popular culture. Because wireless networks are contested, hybrid forms that exhibit commercial, political, and social dynamics on different scales, it makes sense to track itineraries and changes toward wirelessness in terms of an idea (understood in James's sense) such as the "wireless city." Very often the same city has more than one wireless grid overlaid on it. While cities often have a number of different cell phone networks (but rarely more than one sewage or electricity system), these networks usually compete with each for customers by offering similar services that are only separated by spectrum allocations and service contracts. Few people have more than one mobile phone contract. By contrast, wireless networks populate gaps between existing urban infrastructures of transport, communication, energy, and habitation, and can serve very different functions. They are often interstitial rather than infrastructural in the sense of energy, transport, and other grids. The ways they take place, alter space, and modify the daily trajectories of people inside and outside are highly mutable. In some ways their interstitial existence richly epitomizes the thesis of "splintering urbanism" described by Stephen Graham and Simon Marvin (2001, 33):

"A parallel set of processes are under way within which infrastructure networks are being 'unbundled' in ways that help sustain the fragmentation of the social and material fabric of cities. Such a shift, which we label with the umbrella term *splintering urbanism,* requires a reconceptualisation of the relations between infrastructure, services and the contemporary development of cities."

While the splintering-urbanism thesis has many aspects I cannot discuss here, it is not hard to see that wireless networks "unbundle" telecommunications from the twentieth-century public utilities or monopoly service providers. In some ways they offer an extreme case of this unbundling or splintering process. Wireless networks, as they spread along lampposts, corridors, cafés, apartment buildings, hotels, train cars, and parks, easily disappear into the crevices of the architectures, infrastructures, and landforms that Graham and Marvin identify as the fabric of urban life. If "much of the material and technological fabric of cities, then, *is* networked infrastructure" and "at the same time, most of the infrastructural fabric *is* urban 'landscape' of various sorts" (p. 13), then the theoretical problem addressed in this chapter concerns how such interstitial formations intersect with the idea of a wireless city, and conversely how such interstices are imagined and configured in wireless cities.

The presence of wireless networks in the interstices of urban infrastructures, architectures, and landscapes, I will suggest, is deeply entwined with transitions, departures, and arrivals, inflected or steered through *ideas.* Here ideas are understood in a radical empiricist sense as a way of establishing a trajectory or modulating a movement by substituting experiential shortcuts. Through ideas, transitions imperceptibly overlap and associate with each with greater rapidity. An idea is, for James, a substitute movement. Because of the increasingly profuse multiplicity of wireless connections present in many cities, the idea of a wireless city embodies a particularly rich series of continuities and discontinuities. James (1996a, 94–95) writes of space, time, and self: "The things that they [the great continua of time, space, and the self] coenvelope come as separate in some ways and as continuous in others. Some sensations coalesce with some ideas, and others are irreconcilable. Qualities compenetrate one space, or exclude each other from it."

Could this passage be profitably reframed as the time, space, and self of the city? If so, radical empiricism would need to bear on different scales of networked experience in the city and the world in the years 2004–2007, a time in which some things are separate and others "compenetrate." The present chapter explores how wirelessness, with the different layers of conjunctive intensities described in the previous chapter, imbues city places with coalescences and irreconcilabilities concerning arriving and leaving, joining and separating. The two main cities discussed in this chapter are Taipei and London, but many other cities such as Helsinki, Seoul, New York, Stockholm, Berlin, San Francisco, Budapest, Toronto, Sydney, or belatedly, Beijing, also tried to become wireless cities in those years. A later chapter, which concerns transnational wirelessness and development, also concerns urban places, but cities and places such as Lagos, Kigali and Timbuktu in Africa, Everest base camp in Nepal, Dharamsala in India, Bandung in Indonesia, or indeed rural Iowa and northern England.

Following the Lead in London

Walking through city streets has been a standard point of reference for urban sociology since the nineteenth century (Tonkiss 2005). (As I will show, movement in the city is also a guiding thread in James's account of an idea as substitute movement.) The sensation of city walking, although mundane, has often been characterized as overwhelming, chaotic, exciting, dazzling, or enervating. In November 2003, Broadreach Ltd, a UK-based Wi-Fi Internet service provider (ISP), opened a free Wi-Fi "hotzone" centered on Piccadilly Circus, London. The shopping streets of Convent Garden and the entertainment precinct of London's West End meet at Piccadilly Circus. This is a very busy part of the city throughout the day. Broadreach was one of dozens of businesses seeking to capitalize on Wi-Fi networks by offering pay-by-the-minute wireless Internet connections in various public and semipublic places such as railway stations, departure lounges, hotel lobbies, train cars, service stations, McDonald's restaurants, cruise liners, trailer parks, public parks, and cafés. In most cases, these services consisted of a wireless access point providing high-bandwidth

Internet coverage up to several hundred meters for customers carrying a Wi-Fi equipped device (laptop, PDA, VoIP mobile phone, etc.). In contrast to Internet cafés (Wakeford 2003; Miller and Slater 2000), customers use their own equipment and sit or move around according to what they are doing. Since the inception of these hotspots in 2002, the cost and difficulty of logging on to them have meant that they have not been heavily used. As memories of the dot-com crash were still fresh, predictions of another dot-com-style collapse abounded in the IT and business pages of print and online media: "Hopes that the roll-out of wireless broadband networks—so-called wi-fi hotspots—will result in a profits bonanza will be dashed, the technology consultancy Forrester has warned. 'With all the hype today about the rollout of . . . public hotspots, it's as if the dot.com boom and bust never happened,' said technology analyst Lars Godell" (Weber 2003).

Despite or perhaps because of these predictions, many efforts were made to expand the coverage of Wi-Fi hotspots, and to extend them beyond the precincts of the café to include the surrounding streetscape or adjacent open spaces. In Broadreach's case, a new degree of spatial extension was achieved by linking adjacent hotspots located in different cafés, offices, and bars together to form a "hotzone": "The zone stretched from the east end of Piccadilly, from Church Place onwards, through Piccadilly Circus and down Coventry Street as far as Wardour Street," according to the Broadreach press release (Smith 2003). By recent standards of metropolitan wireless mesh, this was fairly modest coverage.

In late November 2003, I sat on the steps of the Shaftesbury Monument Memorial Fountain in Piccadilly Circus with my laptop. These steps are popular with city workers, shoppers, and tourists. In the morning, I had interviewed an artist, Pete Gomes, who was using Wi-Fi networks to broadcast "local television" in the park at Russell Square in Bloomsbury (Gomes 2003). Alongside tourists and lunching city workers, I spent several minutes trying to locate the Broadreach hotzone using my computer. Given that the whole area was well within the hotzone, and that access to the network was supposed to be free until the end of 2003 (Broadreach's CEO Magnus McEwen-King had announced: "In a UK first, we have built this hot zone and we are also offering Wi-Fi access across the UK for free until the end of the year" (Wearden 2003)), it seemed like a good opportunity to see what it would be like to connect to the Internet in the midst of the London buses, the taxis, the barrage of signage, and the flows of people moving toward Leicester Square and Charing Cross Road.

There was no Wi-Fi signal visible. Since I could make no connection, I put the laptop away and walked down toward Leicester Square among a lunchtime crowd of people flowing in both directions, going in and out of shops, cafés, and cinemas. A small dog, a Yorkshire terrier, trotted along ahead of me, weaving between people, occasionally looking back over his/her shoulder. The dog was clearly not a stray or a street dog. It was well fed and groomed. Sometimes it veered off, as if to take a turn up a side street, but then it moved back into the mainstream of people. Sometimes it sped up only to slow down again. Not only couldn't I locate the Wi-Fi hotzone, I was starting to feel responsible for a small dog, lost in the crowd. As I was about to talk to the dog, it looked back over its shoulder toward the other side of the pedestrian zone. Strolling on that side, parallel to me, was a casually dressed man carrying a dog leash. He steered closer to me. In a friendly way he said something like "she knows where she's going."

The coincidence between the unfindable Wi-Fi connection in Piccadilly in November 2003 (surely an experience encountered millions of times every day globally) and an unleashed dog roaming in the crowd is at first glance trifling. But, as we have seen often, for James (1996a, 42), "Any kind of relation experienced must be accounted as 'real' as anything else in the system." Moreover, "real dogs" commonly serve as a model of conceptual experiences for James, just as the pedestrian embodies urban experience for much twentieth-century social theory.[3] What seemed like a potential connection to the Internet without leads or wires, to an experience of the "anywhere, anytime" aspirations of the wireless city of 2003, dangled. It was stuck on the steps of a fountain. However, another experience of connection substituted itself as I walked toward Leicester Square. The lack of a leash and the size of the crowd made any connection between the dog and its walker hard to see. The dog was not on a leash, the man might have explained, because a leash easily gets tangled walking through a crowd. Moving through a crowd with a dog on a leash is harder than moving independently. However, both the wireless networks and the dog entailed trajectories that sought to keep connections on the move. Moving without a lead or leash means establishing connections more often, and sometimes losing existing connections. The fact that there was no connection visible from the steps of the monument that day in November 2003 does not mean connections would never be made there. For radical empiricism, what really exists are not things as such—for example, wireless networks—but transitions, arrivals, and departures.

Experience for James is a trajectory that bundles many connections together, constantly adding and shedding relations, and substituting them conjunctively. Accordingly, a radical urban empiricism in wireless cities of the recent past and near future needs to carefully examine arrivals and departures, especially those from wireless networks to seemingly unrelated substitutes—for example, from hotzones to dog walkers. The basic technique of exploration in radical empiricism, or the operation that makes it empiricist, is to always attend to transitions, and feelings of being in transition, and to rate them as just as real as things, substances, or persons. There are difficulties in doing this. Feelings are not necessarily connected, smooth, or continuous in their transitions. They encounter all kinds of borders and boundaries. While on some occasions, the feeling of transition is not very pronounced, feelings of transition always do somehow accompany transition. James, as Massumi (2002, 6) writes, "made transition and the feeling of self-relation a central preoccupation of his latter-day 'radical' empiricism" (Massumi 2002, 16).

Who in the conjunction of dog, crowds, pedestrians, and hotzone embodies the feeling of self-relation in the transition to wireless networking? Transitions generate sensations because conjunctive relations shift or modulate. As we will see in later chapters, James's insistence on conjunctive relations as the ongoing fabric of experience diverges from most social and cultural theories that tend to cut realities into things, selves, locations, and relations. Nevertheless, a feeling of self-relation associated with this transition may be difficult to detect. Sensations of changing conjunctive relations may not register consciously. A feeling that does not register consciously throws up complications for radical empiricism, but in itself, that difficulty does not mean that such feelings are irrelevant. Feelings of relation that are not large enough to register consciously may still be decisive tendencies. Moreover, feelings that do not commonly consciously register for everyone may well be keenly felt in certain situations or by certain people. _It will be hard to prove that we can work with what we can't see/ aren't aware of_

Transitions and the Feeling of Self-Relation

At the time Broadreach was promoting its unconnectable hotzone, for instance, participants in numerous community wireless network projects such as New York's NYCWireless (NYCWireless 2007) or Montreal's Île Sans Fil (Île Sans Fil 2008) were striving to set up wireless networks in parks, public spaces, and affordable housing. We could say that this is because they have a heightened feeling of potential relations that differs from both the high-profile wireless city enterprises as well as the much more numerous, but static and habituated, uses of commercial wireless hotspots and the coalescence of mobile phone–based network services with Internet services (check e-mail, upload images, browse latest news, etc.) (Forlano 2008). One limit case of their localized challenge to Internet connectivity as the telos of wirelessness is expressed in the work *wifi.Bedouin* by Julian Bleecker (2004). The work was developed in association with NYCWireless and comprises a backpack filled with Wi-Fi access point, laptop, and GPS-equipped PDA. Wearing the backpack, the wifi.Bedouin is not an access point to the Internet, but an ambulatory network (albeit of limited scope) that tends to supplant the Web. Bleecker (2004, 2) writes: "This access point is not the web without wires. Instead, it is its own web, an apparatus that forces one to reconsider and question notions of virtuality, materiality, displacement, proximity and community. WiFi.Bedouin is meant to suggest that what are often considered two entirely separate realms— virtual and physical worlds—are actually a much more entangled hybrid space." In some respects, this description is redolent of the debates over virtuality and physicality that transpired during the late 1990s and early 2000s. However, in other ways, the design and performance of wifi.Bedouin offers the potential to substitute different patterns of connection for the habituated, individualizing, or consumption-based uses of the Internet and Web imagined by the idea of the wireless city.

Furthermore, although a feeling of self-relation does not register immediately, it may register in a mediated form. Feelings of wireless connectivity are nearly always highly mediated. We rely on devices, icons, maps, signs, and memory for those feelings. These mediations can be re-configured. So for instance, certain recent artworks such as Gordan Savicic's (2008) *Constraint City* attempt to render conjunctions between the bodies of pedestrians and nearby wireless networks in cities more palpable. *Constraint City* attempts to create a less mediated feeling. It combines a Wi-Fi enabled game console with a set of motorized straps worn around the artist's torso like a corset, as well as a digital mapping application based on GoogleMaps. In 2007, Savicic wore this equipment as he walked through cities such as Berlin, Vienna, and Rotterdam, cities dotted with wireless access points. As Savicic (2008) writes,

The higher the wireless signal strength of close encrypted networks, the tighter the corset becomes. Closed network points improve the pleasurable play of tight lacing the performer's bustier [sic]. Thus, constituting the aether as a space of possible pregnancy, filled with potential access-points to the networks of communication. Everyday walks between home, work and leisure are recompiled into a schizogeographic pain-map [see figure 2.1] which is fetched from GoogleMaps servers with automated scripts (Savicic 2008).

Such a work is designed to heighten specific feelings of relation associated with urban wirelessness. With its explicit invocation of the work of Michel de Certeau, Charles Baudelaire, and the Marquis de Sade, and its stated aim of making "an urban interface" for an invisible city [of electromagnetic radiation]" based on sadomasochistic fetishes (the corset), *Constraint City* in some ways feels very familiar. The corset tightens in proximity to "close encrypted networks." The description of the work mentions "close encrypted networks" as improving the "pleasurable play of tight lacing." That is, wireless access points that are using encryption trigger the constraints to tighten. The implication here is that encryption has something to do with constraint and control, something inimical to the potential for communication.

However, in *Constraint City*, the feelings of pain as straps tighten in proximity to wireless networks are not shared by spectators, only by the artist himself. We can see that after wearing the corset Savicic's body appears to be marked and bruised. The photographs, the map of the walk, and in fact the whole concept of *Constraint City* together suggest that pain was both welcomed and eschewed. It is not as if encryption represented simply an architecture of control that we should resist. The work can be seen as a pragmatic experiment in constructing ways of sensing relations that cannot be easily shared. Indeed such feelings, feelings that are not obvious to all or widely shared, can perhaps more easily fulfill the function of leading experiences in different directions. They more readily lend themselves to the series of substitutions that James (1996a, 61) argues are of "towering importance" in experience. In the context of the many "wireless cities" that have appeared in the last few years, feelings of relation may well be the object of many attempts to construct, experiment with, and thereby prolong substitutions at different scales. Experience is not, if we follow the radical empiricist trajectory, confined to the conscious experience of a knower. In fact, knower and known are just provisional termini of paths of conjunctively connected experiences. They have no particular ontological status. In any case, transitions comprising experience can remain relatively opaque to the knower either because they are too "large" or not "large" enough to register consciously. Cities stream with conjunctions and disjunctions. However, because they generate so many transitions, cities render it possible to trace transitions and transformations that might not otherwise register consciously. In other words, for a radical empiricist treatment of wirelessness, cities are eminently important domains of investigation.

Figure 2.1
Constraint City (Savicic 2008)

Figure 2.1
(Continued)

Each time a wireless city is announced or launched, we could say, it promises a feeling of a changed connection. These feelings of connection are not abstract or virtual, although they have different charges of empirical potential. Some are closely linked to images and practices of everyday life in the city: commuting, working, studying, dwelling, spectating, socializing, and so on. Many are suffused with broader notions of cultural and knowledge economies. Movements take many different directions, and occur on many scales and termini. Hence, the felt reality of any connection is complex, and encounters many incompatibilities or irreconcilable differences. As the Taipei example shows, experiments in constructing relations can be imbued with many different ambitions. Taipei was blanketed with a Wi-Fi network for the sake of neonationalist technological prestige. Mayor Ma Ying-Jiu's press release has to be put in the context of other announcements of wireless cities in San Francisco, Philadelphia, Berlin, London, Seoul, and Singapore. The construction of wireless networks was very much part of the marketing of global cities in these years. Although Taipei's wireless mesh was announced by the mayor, the network was constructed and run as a commercial service by Q-Ware Communications, and it had the task of actually persuading people in Taipei to register to use the network. Q-Ware describes its service as follows:

Currently, there are nearly (almost) 5000 WIFLY wireless "hotspots" in Taiwan allowing users to access the Internet, including Taipei city's seven main shopping areas, all lines of the underground system (Metro Railway Taipei), 7-11, and chain cafe shops such as Is Coffee and Starbucks café . . . etc., as well as including other chain stores in all cities in Taiwan. The entire WIFLY hotspots in Taipei enables users to get away from the restrictions of cables, access the Internet via wireless at any time and place, and sustain people's working and daily needs. (Qware Communications 2008)[4]

While the access points in the Taipei's WiFly network were numerous (in every 7-Eleven store, in shopping malls, in certain coffee shop chains, in Taipei's MRT subway, as well as in around 5,000 points dotted throughout the city), and offered the possibility of connecting anytime and anywhere to the Internet, they remained largely unused by city inhabitants. Taipei's inhabitants and visitors preferred to use free wireless networks scattered in cafés throughout the city. Perhaps more importantly, there are only a limited number of settings in which wireless network access might be practically relevant to people who nearly all have cell phones, many of them enabled for some kind of Internet access.[5]

Disparate Connections in London

The shape of the city dogs the contours of thought. (Tonkiss 2005, 130)

In the background of all the wireless city projects, enterprises, and experiments lies a common problem: Who is connecting to what? Jean-Luc Nancy (2000, 23) writes that "the city is not primarily 'community,' any more than it is primarily 'public space.' The city is at least as much the bringing to light of being-in-common *as the dis-position* (dispersal and disparity) of the community represented as founded in interiority or transcendence."

If the Internet, and access to the Internet, came to stand in the 1990s for "community . . . founded in interiority or transcendence" (user groups, online communities of many different kinds, etc.), we might say that the wireless city embodies being-in-common as dispersal and disparity. The wireless city was bound to be an unstable, quasi-chaotic entity. Like experience in general, it is not something that simply could be, but a set of tendencies or dispositions. The notion of "wireless network" itself offers sufficiently divergent possibilities to accommodate numerous different desires, practices, embodiments, perceptions, and products. Sometimes wireless networks were intimately entwined with the imaginings of neo-liberal-market-saturated rhetorics of mobility, consumption, and choice through Internet access (e.g., the Broadreach Network, like many other wireless networks, promised that). For instance, if 2003 saw wireless networks extending throughout heavily frequented parts of London such as Piccadilly, in 2004 the Westminster City Council in London announced its "Wireless City" project in terms of social opportunity combined with "real-time" security, public hygiene, and safety service management. In the same parts of the city as the free networker's Wireless London, the City Council project had slightly different aims:

The concept of the Wireless City is potentially one of the most exciting developments in Westminster's history. It will allow us to offer opportunity to our residents through community education schemes on our housing estates and integrated social service provision across the city. We will be better able to reduce the threat and the fear of crime through a flexible approach to community safety, cleansing and CCTV—reacting to events and developments as they happen. It will also help us maintain low taxes through the savings that the scheme can offer. (Westminster City Council 2004)

The mixture of uses here goes well beyond the forms of access to the Internet envisaged in the hotspot model or the alternative infrastructures

envisaged in free networking (described in chapters 3 and 4). At the launch
event in 2004, the bullet-pointed PowerPoint presentation by city information services management focused very directly on city administration:

Order, Opportunity & Low Taxation

• Reduce fixed cost base
• Reduce cost of CCTV extensions
• Reduce installation costs of CCTV and noise monitoring
• Reduce the cost of parking meter management

(Snellgrove and Hearn 2004)

Here, the control systems through which the city government manages
the noise, traffic, crowds, and crime of urban street life became central to
the idea of the wireless city. A year later, in 2005, as the network started
to materialize, its framing directly linked entrepreneurial acumen with
changes in the administrative control of city life. It became the "Wireless
City Partnership." Announcing its partnership with a telecommunications
service provider, BT, and the wireless network equipment makers Intel and
Cisco Systems, the Westminster Council claimed:

This is the first deal of its kind in the UK and will establish Westminster City Council
as a world leader for technology and innovation. The Wireless City will benefit those
who live and work in Westminster by improving the street environment through
reducing crime and disorder, improving the delivery and effectiveness of council
services and enabling us to maintain low tax through delivering significant cost
savings. BT is the ideal partner for us, combining in-depth communications expertise with a strong experience of working with other local authorities to provide
wireless technology. (Westminster City Council 2006)

The wireless city network idea expanded here to include the arrangements and "partnerships" between telecommunications businesses and the
local governments. It also began to encompass a set of relationships
between different entities in the city. Crucially, while the municipality of
Westminster, an important part of central London, was very publicly furnished with a wireless mesh network constructed by Cisco Systems, only
employees of the municipal council could make use of the network.

In the same years, other versions of wireless networks followed in the
wake of the Westminster Wireless City with slightly different motivations.
The Cloud, a UK-based wireless Internet service provider, was granted
access to publicly owned lampposts, benches, and road signs to install
wireless infrastructure for most of the municipal area controlled by the
City of London Corporation. According to the City of London Corporation's press release,

The City of London is the world's leading financial and business centre and has always benefited from a world class communications infrastructure. It is therefore important that we provide the latest technology that will benefit those working in or visiting the City. The Square Mile is a fast-moving, dynamic environment and we are responding to the increasing time pressures faced by City workers by providing the technology for them to stay up to date, wherever they are in the City. (City of London Corporation 2007)

In many ways, this statement echoes Taipei's, only now it is not wireless hardware manufacture, but the pressures of being "the world's leading finance and business center" that justifies the communication infrastructure. As in Taipei, the wireless network in the City of London is run commercially, this time by The Cloud, one of Europe's largest wireless Internet service providers. Taipei's city government made no attempt to justify the network in terms of why anyone would use it. By contrast, the City of London Corporation, like the Westminster City Council, very explicitly nominates "time pressures faced by City workers" (that is, people working in the finance and banking precinct of the City in London) as the justification for the network. So the "world-class communications infrastructure" answers to the needs of finance workers and the demands of the financial transactions they conduct.

Finally, in a slightly different part of London, stretching between central London along the Thames between Millbank and Greenwich, a different kind of wireless network—this time operated by a business branded as "online-4-free.com"—appeared in 2007. The network "gives users free access if they agree to view a 15 to 30 second advert every 15 minutes. If users don't want to view the adverts, they are charged one of a range of tariffs, including £2.95 per hour or £9.95 a month" (Daily Mail Online 2007).

Here the zone of wireless connections along the river sells advertising space to advertisers, and offers them "viewers" in return. Similar "free" wireless networks appeared at the same time in various guises. The failed Google and EarthLink proposal to provide San Francisco with free wireless Internet access relied on a similar mechanism: advertising would be context specific (see Andrejevic 2007).

All of this was taking place against a background of well-publicized "community network" activity in North America, Europe, and Australia. Community or free wireless networks appeared in very many places. In London, *Consume* was a prominent example of an attempt to create a wireless mesh that would cover a whole city. As Priest (2005) writes, "The original idea of *Consume* was to create a metropolitan meshed network that would link users at the edge of the network together into a coherent local infrastructure. This connection would allow collective bargaining for back haul bandwidth, and a free local infrastructure that could support local content and an autonomous media."

The possibility of sharing networks between neighbors in apartments and more densely built-up areas took on multiple significance. Sometimes shared networks were explicitly organized in the form of community wireless networks that sought to enhance access to the Internet for parts of the city that were not so well connected to the Internet (for instance, East London). Community Wi-Fi networks and Wi-Fi user groups could be found in many smaller towns for the same reason. The topology of these networks was different in several ways. As Priest (2005) writes of the ambitions of *Consume*, "This meshed-edge network would provide a challenge to existing telecoms providers by being able to escape from the star topology and its built-in control points. Using an agreement between local network neighbors, the plan was to encourage a systemic de-centralisation and distribution of network ownership and operation." They entailed a different sociality since they were often put together and maintained by groups of volunteers who helped their neighbors join or access a wireless node that itself was connected to a commercial broadband Internet connection. These attempts to build suburb, town, or citywide infrastructures were fragile, sometimes temporary accomplishments. To get information to flow across the network along many different paths, rather than just to and from periphery to center, community wireless groups experimented with different topologies such as meshes and grids. Free networks envisaged a different topology because they sought to do more than connect people to the Internet.

Although they had mixed and partial success, the local, online, and print media visibility of the community networks was important. They became visible as networks just as the free networking groups began to change what they hoped to do. As Priest (2005) says, "The Freenetworking Movement has begun to coalesce around new concerns, with recent discourses putting freenetworks and ownership and control of media infrastructure in a freedom of expression context. Control of a network means ultimate control over network traffic."

As movements, they have been concerned to activate different kinds of political space through wireless networks. Treating information infrastructures themselves as a key political stake, free networking diverges from alternative media or independent media with its focus on independent content or opinion. However, in taking infrastructure as a key site of contestation, these movements have had to engage in activities rarely seen as political.

Community wireless networks were not simple or monodimensional associations between individuals centered around wireless network access in specific locations (such as East London or Greenwich). The case of Wireless London offers an example of the kinds of complicated social organization of wireless networks. The website for Wireless London (2006) claims "Wireless London addresses the creative possibilities, policies, practicalities and potential of Wireless London." The Web site presents Wireless London as a cluster of network topologies informed by many different uses, ideas, alliances, and connections with other organizations. Many of these organizations, such as Freifunk.net, informal.org, CRIT Mumbai (Collective Research), Arts Council England, and hivenetworks, themselves are not strictly speaking part of London. It would be possible to go through each of these partners and trace how its role in Wireless London is defined and understood. The range of interests is very broad. For instance, HiveNetworks "research[es] and develop[s] hiveware for embedded devices and ubiquitous networked computers—tools that enable users to manage space, time and the boundaries around the self in new and previously unthinkable ways" (HiveNetworks 2006).

Any transition to wireless cities is complicated by the fact that many competing associations shape how information moves through cities. As a result, wireless cities are populated by many different kinds of wireless networks on different scales, working on different business models (or not on any business model), and enrolling, registering, and subscribing many different participants. Yet these different networks are not simply incompatible or irreconcilable with each other. "Qualities compenetrate," as James puts it, in urban wirelessness because conjunctive relations effloresce and dissolve around wireless networks as they substitute for each other along trajectories of change. This is a key concern for any radical empiricist approach: it needs to say how the many different transitions that fall under the rubric of wirelessness enter into relation with each at least enough to be part of the sum total of experience. Here, the idea of the wireless city itself can be seen as a thread of conjunctive affiliation. The different implementations of the notion of a wireless city, even within one city such as London, draw attention to this. Each new wireless network acts as a temporary substitute instantiation of the wireless city, and the series of substitutions itself has no pregiven or fixed terminus. Each instance does not simply confirm the wireless city. It extracts something slightly different out of that idea, and thereby changes that idea, so that henceforth, the wireless city cannot be experienced as the same entity. Each instantiation of the wireless city, we could say, substitutes some components in the idea of the wireless city, without ever fully verifying or corroborating it.

Feeling Transition, Moving Conceptually

We can track some of the processes of the feeling of transition to the wireless city by returning to James's own account of what it means to know a thing in thinking of it. James describes this in terms of different itineraries in a city. He describes sitting in his library at 95 Irving Street, Cambridge, Massachusetts, and imagining Memorial Hall, a landmark building at Harvard University: "Suppose me to be sitting here in my library at Cambridge, at ten minutes' walk from 'Memorial Hall,' and to be thinking truly of the latter object" (James, 1996a, 54–55). He asks himself how the mere possession of an idea or even an image of the thing in mind could ever be said to constitute knowledge of the thing. What is interesting and useful in James's answer is his insistence on the role of "special experiences of conjunction" (p. 55) in giving the idea of Memorial Hall its "knowing office." A "special experience of conjunction" could include walking to Memorial Hall along with the reader ("I can lead you to the hall, and tell you of its history" (pp. 55–56)). What is "made" during that imagined walk would be a series of felt transitions that act as intermediaries. The tissue of experiencing these transitions—out of the library onto the street, the street signs, the tower of the hall gradually coming into view—connects the starting point of the knower to the known. The knower—James in his library thinking of Memorial Hall—connects with perception of a thing by undergoing these felt transitions. There is no other way, at least in a radical empiricist account, of knowing a thing.

Now, suppose James sat in a library in Cambridge today thinking of Memorial Hall. He might try to conduct "special experiences of conjunction" through wireless networks. For the imagined ten-minute walk, he might substitute a series of felt transitions to Memorial Hall that went via his laptop through his home wireless network or other available networks in the precinct, accessing web pages, blogs, webcams, and geobrowsers that showed images, directions, maps, descriptions, history, and contact details for Memorial Hall. But that point is fairly obvious. We do not need James to tell us that wireless networks open up different paths for experience to thread along since that is inevitable with networked media. However, in that series of felt transitions from library to Memorial

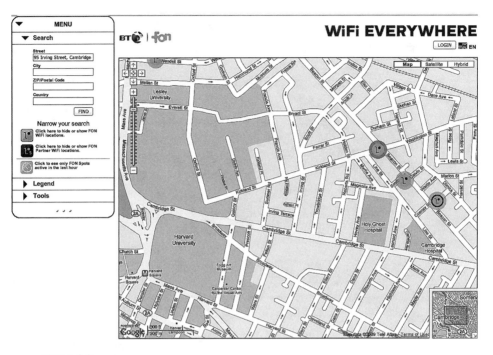

Figure 2.2
95 Irving Street, Cambridge, Massachusetts

Hall, he may well encounter variations in rate and direction. Although it would be impossible for him to be aware of all the intermediary relations and transitions that have to occur for a wireless-mediated knowing of the hall, one question that might come up would be which network to connect to.

He would see, listed on the screen of his iPhone, half a dozen open wireless networks in the vicinity. It could be one of the fifty wireless access points listed by the commercial wireless access-point website jiwire.com as located near Irving Street, Cambridge (see figure 2.2); it could be one of the three wireless access points listed for Irving Street on the website of the "largest WiFi community in the world" (FON 2006). Even after he found the various free, commercial, and community wireless access points accessible from his house, James would need to undergo a series of "felt transitions" as he attempted to access available networks. He might be asked to authenticate himself with a user name and password, he might be asked for network encryption keys (WEP passwords) for the Charles Hotel network, he could be offered the chance to enter credit card numbers

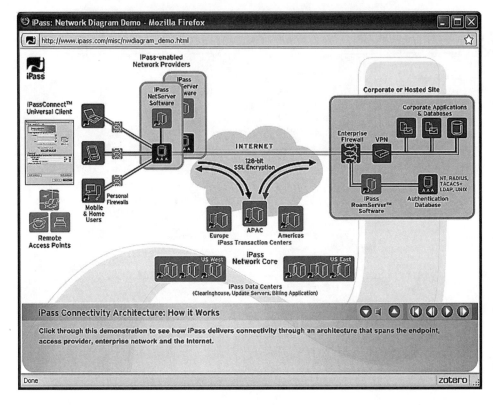

Figure 2.3
iPass wireless network access system (iPass Inc. 2007)

to pay for an hour or a day's connection at the UPS Store or at the Star-
buck's (no. 810) situated 500 meters away. Felt transitions could range
from a familiar, habitual irritation at the demand to enter once again the
same old user name and half-remembered password details, through frustration at not being to connect to a network that "should work," to a
secretive pleasure at gaining access to a network that belongs to someone
else. "Relations are of different degrees of intimacy" (p. 44), he might say
to himself. Supposing James were a business traveler, he might have an
iPass subscription that allowed him to access many networks in his vicinity. The felt transitions would go via something of the architecture shown
in the iPass system (see figure 2.3).

    In short, James's movements toward the known thing, Memorial Hall,
would pass through an externalized series of transitions. In connecting to
any of these networks, the transition from unconnected to connected,
from unassociated to associated, could be felt in many different ways.

This networked variant of the radical empiricist take on what it means to think of something suggests what happens to conjunctive relations today, and hence to the flow of experience under network conditions. There is no experience of convergence, connectivity, or flow that does not go through diverse conjunctive relations, through the transitions that allow "knowing" or "doing" to be felt. These transitions and the feeling of them are crucial to what James calls "nature" or "whatness." Felt transitions are neither spontaneous and random nor completely ordered. The patterns, means, and trajectories of this passing must include variations in rate and direction, otherwise wirelessness as experience of connectivity or convergence disintegrates.

No doubt, the means by which sensations of transition arise are highly complex, and themselves work on multiple scales. But the passing of experience affected by transition can take very circuitous routes. Experience has many different scales, ranging from the impersonal to the personal, from singular to general. On any scale we imagine, wirelessness is not pure flow or pure sensation of transition. It is shot through with temporary termini, with snags and resistances, with circularities and repetitions. Pure wirelessness does not exist any more than pure experience does. Here is James's account of the processing of conjunctive flow through the imposition of stable forms: "Experience now flows as if shot through with adjectives and nouns and prepositions and conjunctions. Its purity is only a relative term, meaning the proportional amount of unverbalized sensation which it still embodies" (p. 94).

We might understand the idea of a wireless city as an attempt to name, fix, and harden the "unverbalized sensation" of transition. Although the conjunctive relations it promises and promotes are on the less intimate end of the scale of conjunction (with, near, beside), they are accompanied by "adjectives and nouns and propositions" that are no less vital to the flow of experience, and that often tend to be much more personal or intimate. What shoots through the flow of experience—"experience now flows as if shot through with . . ."—complicates that flow considerably. This twist or detour in flow is not restricted to wirelessness. The difference between James's imagined walk to Memorial Hall as way of knowing and his accessing wireless networks to find directions and information points to a relatively little attended aspect of experience under network conditions.

The Feeling of Substitution: From Public Space to Safe Space

When municipal or city wireless networks first became popular, they were, like Taipei's WiFly network, imagined as connecting city inhabitants and

visitors to the Internet, to e-mail, and later to telephone-style Internet
services such as Skype. We have already seen that in most cases that reality
did not eventuate evenly. Something mutated or translocated in the course
of the series of transitions toward the wireless city. By 2006, the wireless
city as a bridge for the digital divide had been transmuted into the wireless
city as a safer city, a city with a more secure infrastructure. The relative
failure of the municipal wireless networks as reinvigorated public spaces
did not prevent them from becoming important components in contemporary cities. The more universalist ambitions of the wireless city were
pared down to particular functions and services in the city. Out of the
demise of the Philadelphia EarthLink city wireless network came other
citywide wireless networks with different functions. Tropos Networks, a
North American supplier of metro-wireless networks to cities, promotes
its services under the slogan "wireless broadband you control" (Tropos
Networks 2006), as if to suggest that wireless networks are not in "your
control" but potentially in the hands of unspecified others. It promotes
hopes for expansion in the scope of Wi-Fi in cities: "Wi-Fi's explosive
proliferation gives consumers, businesses and governments a vast and ever-
growing range of fast, easy-to-use and low-cost Wi-Fi-enabled devices. With
more than 120 million units shipped, Wi-Fi isn't just for laptops anymore.
It's also for phones, PDAs, gaming devices, video cameras, parking meters,
utility meters and sensors that detect biological, chemical and radioactive
hazards" (Tropos Networks 2006).

Product manufacturers seeking to supply "metro-scale" networks to
municipal or city government emphasize the heterogeneity of wireless
networks. The product announcements go on to describe how the management of cities can change by virtue of all the different things, places, and
processes that could be interfaced to the networks:

Utility meters, SCADA devices, and parking meters have one thing in common—
they need to be read regularly. This is normally a manually intensive process. For-
tunately the convergence of intelligent, digital meters with networking technology
means they can often now be read from a distance, saving human effort, time and
cost. When coupled with a city-wide MetroMesh network, these devices can now
be read totally automatically, with no human intervention, from one side of the
city to the other. (Tropos Networks 2006)

Urban infrastructure in general becomes part of the wireless network.
The entangling of utilities and wireless networks, or surveillance cameras
and wireless networks, is no coincidence. Cities are inconceivable without
the continuous activity of infrastructural maintenance and repair. As
Steven Graham and Nigel Thrift (2007, 11–12) write, "In many ways, the

inseparable nexus between electricity systems and computer systems presents a particularly useful first example of the massive and largely ignored efforts at continuous repair intrinsic to modern urban life. . . . Beneath the techo-boosterism of the 'new economy,' the realities of using contemporary computer systems is, in many ways, constituted through continuous repair and maintenance."

Hence, utilities such as energy companies have a strong interest in becoming wireless network service operators, since the work of maintaining an energy infrastructure increasingly calls for control operations at all points, not just in the control rooms. Reading meters, monitoring malfunctions, and detecting breakdowns can all be done remotely if the infrastructure or energy grid itself is overlaid with a wireless network. So, for instance, in early 2006 the main electricity supplier to the city of Toronto, Toronto Hydro Telecom Corporation, began installing a citywide municipal wireless network called "OneZone" (Tyler Hamilton 2006). This was not so much because it wants to compete with telecommunications companies as such but because "smart meters" help automate meter reading,

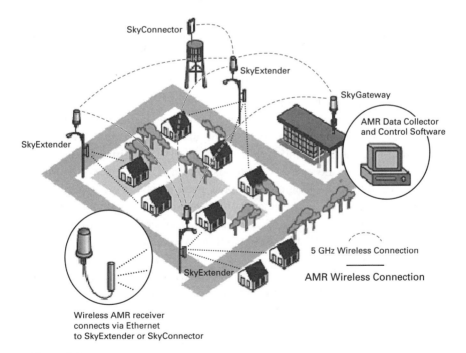

Figure 2.4
SkyPilot city wireless administration (SkyPilot Networks 2009)

monitoring, and billing for the electricity infrastructure, as well as providing subscription-based Internet access (Clement and Potter 2008).

City administration often hope wireless networks will be self-organizing and self-administering, even as they install them in order to simplify, automate, or reorganize administration and maintenance of other urban infrastructures (such as surveillance, transport, and energy) that stand in constant need of repair and upkeep. The briefing for Wireless Philadelphia claimed in 2006 that "once the Wi-Fi units have been installed, they create a self-organizing and self-healing wireless mesh. While some of the units will require a high capacity connection to the Internet, most units will only require access to a power source that can be readily obtained from the existing street or traffic light" (Wireless Philadelphia Executive 2006). The idea of the city mesh network as "self-organizing and self-healing" is an important development in the fabric of the wireless city. It promises to reduce the technical work needed to keep the city running.

Finally, "the people of the city" appear in the idea of wireless city. For instance, the Executive Committee of Wireless Philadelphia invokes a version of the digital divide to help justify becoming a wireless network provider: "A wireless city will be a strategic investment in the people of the city. It will provide an infrastructure that can assist in bridging the digital divide that now exists and prevents many individuals and families from obtaining the full measure of the opportunities generated by the Internet because they can't afford the cost of wired broadband Internet access" (Wireless Philadelphia Executive 2006).

Perhaps more of an afterthought or retrospective justification that capitalizes on the visibility of free and community-access wireless networking, the municipal networks incorporate the "digital divide" into the network image. The experiments and inventions associated with the community-access wireless networking projects have given a public visibility to Wi-Fi that commercial product releases cannot achieve without significant cost. Just as the invocation of the "digital divide" vindicates important changes in the control, security, and administration systems of the city, so too has the wireless network allowed cities to make themselves visible as centers of innovation and creativity in which infrastructures, work, and lifestyle are closely linked: "Providing wireless accessibility will be as important in the future as traditional utilities. Providing an environment that attracts and keeps the knowledge workers who drive the economy of today is all-important. The economic development benefits of this strategy are clear and compelling. Not only will it make Philadelphia a better choice for firms to do business, but it will enhance the desirability of the city's neighborhoods as a place to live" (Wireless Philadelphia Executive 2006).

The image of the wireless network became a metasymbol for hundreds of cities, "enhanc[ing] the desirability of the city's neighborhoods." In this image, a wireless flow of images, texts, voices, transactions, and media content filled the neighborhoods where "knowledge workers" live.

Unstable Wireless Cities: Experiments in Substitution

The idea of a wireless city cannot stabilize a homogeneous space of communication.[6] This is partly due to very material problems of signal propagation. It is also, as we will see in later chapters, because the distribution of wireless connections is not always easy to visualize or map. At a technical level, as a book on wireless network hacks observes, "Perhaps the most difficult task in wireless networking is trying to visualize what is really going on" (Flickenger 2003, 43). (Hence, various attempts have been made to produce citywide images of 802.11 Wi-Fi networks in particular cities: "In April 2004, Humphrey Cheung and some colleagues flew two single-engine airplanes over metropolitan Los Angeles with two wireless laptops. The project logged more than 4,500 wireless networks" (Marriot 2006).) Furthermore, with the "splintering-urbanism" thesis, one of the most notable features of city wireless networks has been the difficulty of imaging what scale they operated on, where they were located, and who was connected to them. In contrast to the centrally planned, diagrammed, and managed infrastructures of landline and mobile telephones, electricity, water, and sewage, wireless networks operate on many different scales, ranging outward from domestic interiors, through neighborhoods, suburbs or town centers, citywide meshes or networks, and occasionally across wide geographic distances (up to hundreds of miles in experimental setups). In any case, urban and nonurban wirelessness are inseparable. The same technology, which was originally envisaged as a way of networking offices and houses, can also now be found in startlingly many locations, in trains, vineyards, cattle ranches (Biever 2004), wildlife preserves (Cohn 2004) mountaineer base camps, as well as cafés, hotels, libraries, airports, and living rooms (Mackenzie 2005). At a hotspot, the same access point (AP) may be accessed by traveling businesspeople doing e-mail, or by neighborhood residents sitting in cafés. In short, there are many ways of siting and adapting wireless networks, including some that range outside and between cities. Indeed they often represent ways of overcoming some gap or difference between life in the city and life outside it.

Where then does the idea of a wireless city terminate? Perhaps in no place or thing. There may not be any actual jumping, barking, hairy dog to be found. In contrast to much urban sociology influenced either by European or American thought in which terms such as *community*, *diversity*, *neighborhood*, *place*, and *space* loom large (Tonkiss 2005), the substitutions progressively attached to the idea of the wireless city hold quite different scales of practice and connection together in series. The wireless city as advantageous pathway—James's basic notion of an idea—comprises a series of substitutions, alternate developments, failures, and unexpected outgrowths in wireless connections to information networks. The substitutional series cannot be easily divided into macrolevel scales and practical, performative engagements with urban spaces. The idea of the wireless city does not stand still, or arrive at a particular place. The wireless city brings together a changing media formation, the Internet, and the open, porous, mobile, and shifting forms of cities. In many ways, the wireless city is a bland substitute for the often intense sensory charge of urban environments. Ideas can "make the flux impossible to understand," James (1911, 79) writes. Practically, however, if ideas are trajectories that entail corroboration and continuation, even relatively empty ideas can affect the relations in their environment. They can allow substitutional series of transitions to unfold without necessarily terminating in embodied perceptions or consciously registered relations. Perhaps, like wirelessness more generally, the idea of the wireless city terminates in a movement, not in a place.

3 Wireless Chips: Digital Signal Processing as Conjunctive Envelope

Never have so many, so diverse and such demented operations been multiplied in order to draw from the depths an intensive *spatium* a serene and docile extensity, and to dispel a Difference which subsists in itself even when it is cancelled outside itself. (Deleuze 2001, 234)

In short, in the world constituted by electromagnetic cosmology (and industry), understanding the electromagnetic field is the only way to understand ourselves and our surroundings. (Bureau d'Etudes 2007)

At the end of 2007, one billion IEEE 802.11 or Wi-Fi® chipsets were in the world. One billion such chipsets will be produced *each year* by 2012, according to market researchers (ABIResearch 2007). Most of these will not go into computers. Two-thirds will find their way into electronic devices, especially consumer electronics and telephones, and many will percolate into wireless network infrastructures in cities, in industrial and institutional facilities, and in environmental sensor networks. Recent market projections envisage a trillion wireless devices in use by 2015. This is called "teraplay" (Gabriel 2009). In terms of sheer scale of communication networks, never have so many, and so diverse, operations been multiplied. What kind of assemblage is in the making here? What kinds of feelings or sensations do these numbers scale to?

We can imagine the architecture of wireless chipsets as a site of intense conjunction. This architecture is daunting in its technical and mathematical complexity. However, the apotropaic effect of technical architectures presents, in James's terms, a problem to be explored, not shunned. The key analytical problem is to find a path through this technical complexity that allows us to sense how it allows worlds to "hang together." In wireless devices, chipsets perform digital signal processing (DSP) on radio-frequency signals. As wireless chipsets multiply, they envelop an indefinite number of processes that alter sensations of location and situation. I will argue that

DSP intensifies conjunctive relations, and heavily augments arrivals and departures. By combining a number of different algorithmic processes, they generate a conjunctive envelope, a spatial-temporal fold that configures and concentrates arrivals and departures. In consequence, wherever this envelope appears, and particularly so in cities, the way a world hangs together is altered. On the one hand, wireless chips extend and ramify digital networks in ways that seem predictable and quasi-inevitable. On the other hand, these alterations are not reducible to or fully captured by the figure of the network. They are perhaps better expressed in terms of pedestrians on city streets than with respect to the flows of information between nodes in networks. The "serene and docile" extension of wireless networks, I will suggest, draws from the "intensive spatium" of the city and its many conjunctive relations. Quasi-structural aspects of wireless DSP come into tension with the figure of network as a regulated flow of messages. As we will see, the wireless signal presents a domain of excess pathways and overwhelming openness to cross-signaling, multiple paths, loss of connection, and interference from others. As in other chapters of this book, I concentrate on 802.11 or Wi-Fi chipsets.[1] However, across the proliferation of different wireless standards (Bluetooth, Wi-Fi a/b/g/n, WiMax, ZigBee, GSM, CDMA, 3G, LTE, TD-SCDMA, etc.), very similar DSP techniques are used.[2]

Spark Gaps: From Wireless 1900 to Wireless 2000

There are groups of people who know an extraordinary amount about wireless chipsets and whose experience of them must, in principle, be detailed and varied. Chip designers and communication engineers would the main example. Design work on wireless chipsets is a highly distributed, multilayered, and globalized activity, and study of the design processes would require multisited fieldwork across engineering laboratories, factories, software and design tool development, trade and academic conferences, thousands of patents, very lengthy standards documents (the 2007 IEEE standard for 802.11 WLAN runs to 1,232 pages), and so on. Another much smaller group worth studying would be the people who write open-source software drivers for specific commercial chipsets. Although their concerns do not usually relate to the actual signal processing, they also have a well-developed sense of differences in speed, power, timing, and so forth associated with wireless DSP. Ultimately, however, anyone using gadgets like laptop computers or mobile phones also has some degree of awareness of chipsets. Sensations of wireless connection are many and varied, and they heighten and diminish across different settings and circumstances. Different chipsets configure these sensations differently.

Whatever turns and twists in experience or feeling of wirelessness we examine, contemporary wirelessness is framed by a large-scale communication shift to the medium of electromagnetic waves. This is itself based on Maxwellian physics that dates from around 150 years ago, and its unification of electrical, magnetic, and optical phenomena in the key concept of the electromagnetic field (Maxwell 1865). By 1900, many different relations of connection, transfer, proximity, and influence were imagined around electromagnetic or wireless waves. As Sungook Hong (2001, 2) argues in his history of wireless invention, there was no single inventor of wireless communication because there was no single problem. Wireless telegraphy was a key commercial application, but wireless communication was also a scientific challenge, a way of experimentally vindicating Maxwellian physics, and an important component of military, colonial empire. Many people were trying to invent wireless communication for different reasons in Britain, Russia, India, Germany, and the United States (Marconi in Italy/the United Kingdom, Tesla in the United States, Braun in Germany, Popov in Russia, Bose in India, etc.). There were some common elements to the problem, such as how to create a continuous signal (a periodic waveform), how to propagate it effectively (antenna design), and how to pick up and amplify the signal at a distance. But the significance of wireless communication was not clear. In contrast to wired media, as the 2007 IEEE 802.11 standard states, wireless is "a medium that has neither absolute nor readily observable boundaries" (IEEE 2007, 23). Electromagnetic oscillation is an event whose scope could be measured by the multiplicity of its interpretations, as Isabelle Stengers (2000, 67) would say.[3]

Philosophical understandings of experience were oscillating too. Phenomenology was questioning the givenness of time and space as a geometric continuum, psychoanalysis was questioning the primacy of conscious thought, and philosophers such as Henri Bergson, William James, and John Dewey were rethinking the relation between thought and action, thinking and things. In particular, James's radical empiricism was dismantling any pregiven distinction between inner and outer, between experience and thing. For him, relations exist just as much as things. Indeed, James has a very general notion of experience that easily embraces things and thought. Thinking and things are not very far apart for him: "*thoughts in the concrete are made of the same stuff as things* are" (James 1996a, 37). James expands experience to envelop diverse "inner" and "outer" processes: "To be radical, an empiricism must neither admit into its constructions any element that is not directly experienced, nor exclude from them any element that is directly experienced. For such a philosophy, *the relations that connect experiences must themselves be experienced relations, and any kind of relation experienced must be accounted as 'real' as anything else in the system"* (p. 42).

This rule of always including relations between experiences, and any experience of relation, almost defines the core of radical empiricism as a technique of philosophically reconstructing experience. It also allows James's work to be read as a philosophy of movement and mobility in several ways that I find interesting and relevant to the problem of "why so many chips? why so much DSP?"

For an analysis of wireless signal processing as experienced relation, two facets of radical empiricism are immediately relevant. First, James often uses wave- or waveform-based understandings of experience to expand the envelope of relations comprising experience: "We live, as it were, upon the front edge of an advancing wave-crest, and our sense of a determinate direction in falling forward is all we cover of the future of our path. It is as if a differential quotient should be conscious and treat itself as an adequate substitute for a traced-out curve. Our experience, *inter alia,* is of variations of rate and of direction, and lives in these transitions more than in the journey's end. The experiences of tendency are sufficient to act upon" (p. 69).

This is a well-known passage from James. The problem is that James is easily read as if he is speaking as a psychologist rather than as a philosopher, or as a metaphysician rather as an empiricist, as if this predominance of tendency was an attribute of an experience interior to a subject. The wave crest is usually understood as something that rolls across the mind, not outside it, not in the world. We could say, however, that early twentieth-century wireless cultures begin to channel that wave crest of experience in multiple directions. In the state of wirelessness, experiences of tendency or of advancing wave-crest situations become just as important as things or thoughts.

There is a second aspect of James's radical empiricism that lends itself to wirelessness, and more immediately, offers a way of thinking about what is happening in the proliferation of signal processing. By virtue of the primacy it adamantly attributes to conjunctive relations, radical empiricism closely follows the paths that connect parts of a world:

Radical empiricism takes conjunctive relations at their face value, holding them to be as real as the terms united by them. The world it represents as a collection, some parts of which are conjunctively and others disjunctively related. Two parts, themselves disjoined, may nevertheless hang together by intermediaries with which they are severally connected, and the whole world may hang together similarly, inasmuch as *some* path of conjunctive translation by which to pass from one of its parts to another may always be discernible. Such determinately various hanging-together may be called *concatenated* union. (p. 107)

What he describes here, particularly the way the world "hangs together" as concatenated union through conjunctive paths, could be seen as a philosophically couched description of the growth of wirelessness on many scales. As we will see in chapter 7, even a "world" for James is a set of connected conjunctive relations ("concatenated union").

What is "radical" here is the pragmatic importance given to *conjunctive relations*. In philosophical logic, James argues, conjunctive relations such as "next to," "in," "between," "beneath," and "behind" are normally seen as less accidental or vestigial in comparison to the disjunctive relations of difference and distinction that define what really exists. By revaluing conjunctive relations, James offers an extremely lightly structured way of accounting for the streamlike nature of experience, prior to any opposition or sorting of experience as "inner" or "outer," "thought" or "feeling," "doing" or "thinking." Relations of proximity, intimacy, availability, and separation take on much greater potency in radical empiricism. Linked together ("concatenated") in sets ("union"), they make worlds. Hence, radical empiricism offers a way of following networks and wires, wiring and unwiring, without subordinating them to the figure of the network.

The Problem of Connection: So Many Signals, So Much Interference

Guglielmo Marconi claimed that he received the three letters "SSS" transmitted from Poldhu in Cornwall, England, at St. John's, Newfoundland, on December 12–13, 1901 (Hong 2001, 54–55). Some scholars today argue that he may well have mistaken atmospheric noise for a Morse code message. The immense apparatus at Poldhu emitted quite powerful, chaotic or "dirty" electromagnetic discharges (25 kilowatts; Aitken 1985, 265). By today's regulatory standards, they would certainly be illegal. Jumping a century from Marconi and Tesla's wireless telegraphy to wireless networks, we are confronted with a very different social-technical-architectural-corporeal assemblage. Although the antennae, even the amplifiers, are similar in principle (but not scale), the algorithmic complexity of wireless networks looks very different from the Morse code Marconi and Tesla transmitted. Today, in 2009, it is just possible to use a wireless network while flying across the Atlantic. Various airlines have been offering this service for the last five years. But wireless networks are much more densely clustered in cities and urban places. What has happened in the last century to radio signals?

Today, billions of chipsets wind a tight knot of DSP around radio antennae. Wireless chipsets, produced by Broadcom, Intel, Texas Instruments, or Airgo, are tiny (<1 cm^2) fragments that cram highly convoluted and concatenated paths into the small "form factors" associated with wireless devices and gadgets such as Wi-Fi USB sticks, Wi-Fi/3G mobile phones, or Wi-Fi-enabled SD memory cards for cameras. From a radical empiricist perspective, Wi-Fi, Bluetooth, and mobile phones (but also including DAB, DVB, etc.), with their billions of miniaturized chipsets, proliferate conjunctive relations. Today, in the wake of the dispersion of electronic broadcast media into digital communications and information networks, the paths of translation that make up the world are incredibly numerous, and inevitably very entwined with each other. As we have already seen in the previous chapter, urban life and contemporary patterns of global mobility travel generate constant conjunctions, concatenations, collections, and other forms of "hanging together." Cities in particular are conjunctive as well as disjunctive places. Urban and media experiences are loaded with "variations in rate and direction" associated with conjunctively processed paths.

This proliferation of paths and conjunctions responds to the relational problems of cities. As information and messages move around more on many different scales, people move more too. The global movement of images, messages, and data, and the movement of people, are closely linked (a point made by Appadurai (1996) well over a decade ago). As communication intensified during the twentieth century, the mobility of people and the signals they transmitted and received became more densely interwoven and mutually transformative. As patterns of movement in cities, between cities, and between regions, countries, continents, and hemispheres expanded (for so many different reasons), they took communication on the move with them. Hence flows of information multiplied, and networks proliferated around those flows. Networks such as the Internet, but also telephones, became more dense, and imbricated lived spaces. The infrastructural problems of putting wires and cables everywhere have been increasingly albeit partially solved by multiplying radio-frequency waves.

While the notion of wireless networks implies that there are fewer wires, it could easily be argued that actually there are more wires. Rather than wireless cities or wireless networks, it might be more accurate to speak of the rewiring of cities through the highly reconfigurable paths of chipsets.

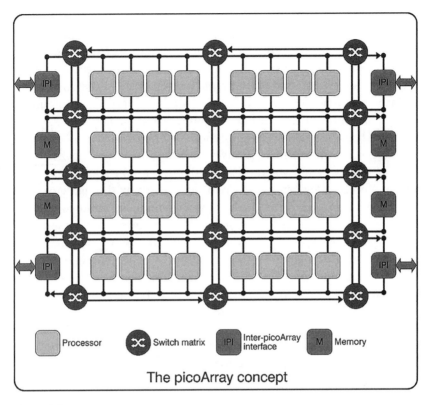

Figure 3.1

Typical contemporary wireless infrastructure DSP chip architecture PicoChip202.
Credit: PicoChip (PicoChip 2007)

arch of cities
arch of DSP chips

Billions of chipsets means trillions of wires or conductors on a microscopic
scale. The "architectures" of chipsets resemble cities viewed from above
precisely because they internalize many of the relational processes of
movement in cities. In contrast to the cables and fibers of the Internet or
standard telephone networks, these wires can be configured and connected
using the different levels of abstraction present in software and code. The
PicoChip (figure 3.1), a DSP chip for wireless communications with an
array of hundreds of programmable processors, epitomizes how such connections become an ongoing object of intense focus. Wireless signal processing concentrates in the intensive spatium of the chip what other
infrastructures accomplished with roads, pipes, cables, control centers, and
ducts. The intensive spatium of signal processing is the core concern in
telecommunications today.

The sheer density of wireless transmissions creates fresh problems of regulation, interference, competition, and overload. The problems of interference, of unrelated relations, are potentially immense. On the one hand, governments and states control spectrum allocation to prevent interference between civilian and military communications. On the other hand, civilian spectrum is crowded all around with mass media, organizations, groups, and individuals transmitting and receiving signals. Spectrum has therefore become a valuable, tightly controlled resource. For any one communication, not much space seems to be available. And even when there is space, it may be noisy and packed with other people and things trying to communicate. Signals may have to work their way through crowds of other signals to reach the desired receiver. Communication does not take place in open, uncluttered space. It takes place in messy configurations of buildings, things, and people, which obstruct waves and bounce signals around. The same signal may be received many times through different echoes ("multipath echo"). Hence the two "disjoined parts" James refers to might often come into relation through very complicated paths of "conjunctive translation." Waves are so prone to additive synthesis. This is the basic problem of wirelessness today: How to establish intermediate paths amid so many possible conjunctive relations without experience dissipating into pure noise? Because of the presence of crowds of other signals, and the limited spectrum available for any one transmission, wirelessness needs to be very careful in its selection of paths if experience is to stream rather than just buzz, as it may have done for Marconi in 1901. In contrast to the early twentieth century, the problem is not how to jump a gap (the Atlantic), but how to differentiate amid so many converging paths.

Air as Algorithmic Envelope

The *air interface* is a term for that part of a wireless or mobile telephone network that lies between the antennae of a device and a base station. It is an elusive interface, one that shows no face apart from the tips of antennae and the more or less conspicuous towers and masts of telecommunications and telephone service providers. The air interface, however, is synthesized by technical processes expressed in algorithms. These algorithms generate waveforms that support conjunctive pathways. DSP algorithms are often regarded as complicated mathematical forms of abstraction or mechanism. If acknowledged at all, they are regularly treated as the most abstract aspect of electronic media and communication technologies, the part that lies closest to mathematics. We need a much more sensitive treatment of their becomings. They transduce diverse realities. In analyzing air interfaces, we need to elicit what in wireless communication is not reducible to "mere calculation" because it multiplies calculation to such an extent that it can longer be regarded as "mere." They reorganize movement and flow in very specific, and I will suggest, composite ways.[4] In the wake of James's radical empiricism, and sociological work on postsocial relationalities in markets and sciences (Knorr-Cetina and Bruegger 2002; Latour 1999), we can regard them as ways in which worlds can be made to hang together, at least partially. As a point of departure, we should not assume that algorithms are either objects or subjects, or even intersubjective relations. Rather they are relational situations concerned with transitions between states. They are marked by much thought or mental effort (in the form of design, modeling, experimentation, and simulation); they operate physically, amid states of affairs (institutions, residences, markets, public spaces, transport, etc.); and they connect movements to each other (in the form of streams of communication—conversation, collections of textual and graphic elements, image flows, sensor data, control data). Algorithms cast a certain shadowy subject position (embodied principally in the figure of the electronics engineer, sometimes in a proper noun such as "Viterbi decoding"), and they display certain collective attributes (for instance, by being the object of intensive efforts of standardization, optimization, and the target of intellectual property claims), but something in their architecture creates unstable movement between "it" and "we," between thing and thought.[5]

How can radical empiricism work with something like DSP algorithms? They seem far from the felt realities and direct experience James discusses. However, I would argue that the very architecture of the chips and the algorithmic traits of wireless signal processing actually owe much to experiences of moving through cities, of being surrounded by noise, of finding the shortest route, and of forming constant impressions of potential degrees of intimacy and distance from others. No doubt, the chipsets under discussion here, such as the PicoChip "PC202 Integrated CPE/Access Point PHY/MAC Processor" (PicoChip 2007) or the Broadcom "BCM4325 Low-Power 802.11a/b/g with Bluetooth® 2.1 + EDR and FM" (Broadcom Corporation 2007), are somewhat opaque to analysis because their "making" stretches across university research projects and publications, sophisticated Electronic Design Automation (EDA) software, many simulations and tests, complicated technical standards (such as IEEE802.11), convoluted intellectual property portfolios, and chip "fabs" or foundries scattered across East Asia, North America, and Europe.[6] The chips, more-over, with all their algorithmic-architectural density, hardly appear in the everyday world at all. Even though they are made in the millions (and soon, according to numerous market projections, billions or trillions), they quickly disappear into consumer devices and infrastructures. They are rela-tively mature, noncontroversial albeit mutable facts. They "work" (mostly: see the discussion of "bricking" in the next chapter). Yet if we regard the chipsets as practically given, how do we have any experience of what goes on in or around them? As Massumi (2002, 16) suggests, "A complication for radical empiricism is that the feeling of the relation may very well not be 'large' enough to register consciously."

It is reasonably uncontentious to say that wirelessness as a mode of contemporary experience depends on an algorithmic mosaic of calcula-tions carried out to allow communication to occur in the presence of many others. Saying this, however, does not mean that we have any experience of algorithms as relational forms, or as architectures for conjunctive trans-lation. What awareness, even marginal awareness of algorithmic complex-ity of wireless, is possible when confronted with such a state of things? In this chapter, my treatment of algorithms as generating a conjunctive enve-lope is partly guided by Gilles Deleuze and Felix Guattari's (1994, 117) account of the relation between philosophical problems and scientific matters of fact: "When an object—a geometrical space for example—is scientifically constructed by functions, its philosophical concept, which is by no means given in the function, must still be discovered."

This injunction to discover the philosophical concept from the states of affairs referenced in scientific functions is somewhat onerous.[7] However, James offers a relatively slender yet strong way of doing that. Exploring wireless algorithms as conjunctive relations tells us nothing about what it means to actually live wirelessly in some city. Yet algorithms already carry many of the tendencies described in James's philosophy of overabundance, and, as we will see, the algorithms themselves in important ways express awareness of contemporary urban life. Hence to treat the loops and knots of wireless network DSP as conjunctive means nothing more than to locate the "objective distribution of the singular and ordinary" that defines it (Delanda 2002, 116). "Objective" here, I take in a pragmatist sense: a quality is "objective" if "it in some way reacts to its neighbors" (Redding 2001, 259). This search for something "objective" in wireless algorithms does not aspire to produce a higher knowledge of signal processing, or a transcendental philosophy of wirelessness. Rather, the view of wireless signal processing as a "conjunctive envelope" attempts to convey something of the forms of divergence inhabiting the event of wirelessness, as it is frenetically interpreted in wireless gadgets, systems, infrastructures, and networks.

If we do not appreciate the work of algorithmic processes, then we cannot understand the effervescences associated with wirelessness—how chips multiply and flow in their billions into markets, how new forms of embodied habit and conduct appear, how places are transformed into zones of connectivity, and how new species of sociality concresce around them. We also can't understand the growth of wirelessness, its proliferation, its overlaps and contestation, or its structuring of collective experience. Wirelessness as a state of effervescence develops in assemblages of conjunctive relations: it lies at the fringes of experience, but tinges experience with certain feelings of proximity and attentiveness that may very well not register consciously.

How to Make a World Hang Together: Noise

People in cities are closely adjacent or distant, moving in concert or jostling each other, stationary or in transit, inside or outside. Their sense of living, of circulating between places no matter how far flung or local, comes from the conjunctive relations that put their bodies with, nearby, and against other bodies. "'Body' really means what is outside, insofar as it is outside, next to, against, nearby, with a(n) (other) body," Jean-Luc Nancy (2000, 84) writes. The flow of experience today circulates through algorithmic processes that stream conjunctive relationality between bodies.

It is not too hard to understand an algorithm as something that brings things into relation according to a certain mathematical or functional order. That would fit directly with the description of algorithms found in DSP textbooks. But how do we orient ourselves to the ways they are experienced, often at the limits or fringes of perception? The DSP algorithms are not like credit card transactions or the sorting functions in spreadsheets. They have a much greater investment in the prevailing conditions in heavily populated, device-congested, urban environments. Rather surprisingly, contemporary wireless signal processing tries to generate signals that look as much like noise as possible. Noise, which we normally associate with the presence of others and think of as parasitic or disturbing, is the basis of an increased density of communication as well as being a way of managing interference in contemporary digital signal processing. The signal-processing algorithms used in wireless networks such as Wi-Fi, Bluetooth, and 3G are not unique in their tendency to create signals that almost pass as noise. The techniques of multiplexing, transformation, compression, and error correction are found in many places—in audiovisual digital media, scientific computing, many forms of communication, and signal processing. However, the algorithms working in a wireless network such as IEEE 802.11a (IEEE 1999) or 802.11g (IEEE 2003) form a tightly woven fabric of different processes. The ways these different processes fit together are important. Their interwoven texture creates an envelope that allows data to circulate in the crowded signal channels of urban-electronic space as if it were just noise.

I focus here on signal processing occurring in the so-called physical layer (PHY) of information networks, rather than the addressing and routing that occurs at higher layers in the communication protocols (for instance, where the Internet proper appears). Most of this physical layer or air interface processing is quasi-hardwired into semiconductor chips, although these chips themselves increasingly support ongoing reconfiguration. The Physical Layer is that part of network infrastructures that has most to do with lived spaces, and yet it is not usually seen on screen. It grapples with cables, antennae, plugs, sockets, satellites, and other forms of physical constraint.

It is difficult to imagine a diagram that could convey how the different algorithms work together to effect communication in the physical layer. At most, typical signal-processing diagrams show a succession of steps represented by boxes (see figure 3.2). A typical digital signal-processing textbook would take each of these boxes in isolation and describe what is happening in it (Boccuzzi 2008). The set of algorithms in Wi-Fi (specifically, 802.11a and 802.11g) sometimes give the impression of being a discrete sequence of operations on data to be communicated. However, that impression is misleading. The overall conjunction of encoding-transmitting-receiving-decoding wireless communications is much more interwoven than a simple series of boxes can convey. The juxtaposition of different components constructs a signal envelope or composite waveform that is open in certain ways and heavily closed in other ways. Information is coded in a sequence of steps, but these steps take account of each other. Information is encoded a number of times to allow different relations to be entwined with other. It would be misleading to think of wireless communication as simply transporting packets of information via radio waves. What animates this movement is a complicated set of conjunctive relations between different parts of the signal.

Two components of the conjunctive envelope, two of the algorithms involved in the construction of the signal, can usefully exemplify the

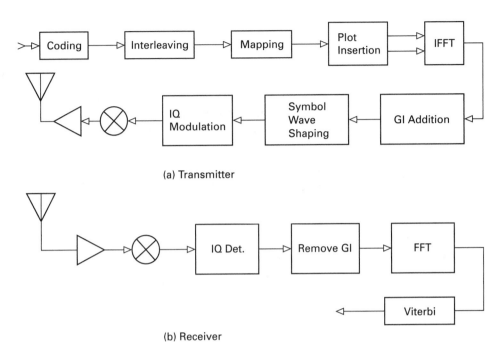

(a) Transmitter

(b) Receiver

Figure 3.2
Concatenated algorithms in wireless computation

texture of the conjunctive envelope. Their modes of operation are very different from each other. The fast Fourier transform (FFT) stages of transmission and reception, and the convolutional coding–Viterbi decoding stages (see figure 3.2), are respectively the most "outer" and "inner" parts of the algorithms in 802.11a/b networks. These are the zones of the conjunctive envelope that lie closest to the antennae and closest to the sources of digital information or communication. Together they highlight two primary aspects of experience in James's account: the wavelike tendencies that express variations in rate and direction amid transitions, and the "concatenated union" that allows worlds to "hang together" amid disjunction and disconnection.

Transformations and Tendencies

What can be experienced of the algorithmic character of signal processing in contemporary wireless networks? Even if it is barely perceived because it is so technologically sedimented, DSP nonetheless structures feelings of relationality and even self-feeling. Everyone experiences something of this whenever they use a mobile phone or other digital wireless devices in a particular place. The slight differences in availability, celerity, clarity, and responsiveness of contemporary wireless signal environments, as well as their breakdowns, exclusivity, and unavailability, emanate in part from the way signal processing compensates for and exacerbates the conjunctive nexus of a particular setting, be it urban or remote. These sensations of clarity and availability soon become mundane. Hence, one tendency of contemporary wireless media is to constantly increase the rate at which signals transport data. The conjunctive envelope frequently expands to include new relations: VoIP, streaming video, virtual private networks that cross many network boundaries, or so-called push e-mail (as in the BlackBerry service or in recently developed applications for the Apple iPhone). Much of this expansion relies on compressing a sequence of movements into ever shorter time intervals. It is not specific to wireless networks since a similar conjunctive envelope appears in digital television and digital radio.

How does wireless DSP create a path between two potentially mobile points in a city? This problem, as I have already said, elicits a variety of engineering responses. The designs of specific wireless chipsets (and indeed different wireless standards) represent different attempts to find an optimum balance between speed and reliability, signal strength, and power consumption, in the midst of the unbounded yet cluttered, unprotected yet security-concerned medium of air. The designers of contemporary wireless DSP chipsets usually supply a palette of different hardwired algorithms alongside generic processors. Communications engineers can programmatically combine these in different ways for particular applications. The PicoChip, for instance, supplies "several hundred heterogeneous processors" (Panesar et al. 2005, 323) as well as specialized "accelerator units" for specific algorithmic processes. In almost all contemporary wireless network DSP, some version of the fast Fourier transform appears. Late in the 802.11b Wi-Fi processing train, when a signal is close to being transmitted, and earlier in the decoding process, the FFT and the complementary inverse fast Fourier transform (IFFT) appear in order to perform "multicarrier modulation," a technique that divides data across a number of channels that carry signals in parallel. This technique allows data to move more slowly, and with less susceptibility to noise and interference. Indeed, as we will see, the design objective of the algorithm is to render the signal itself as much like noise as possible. The FFT and IFFT are critical in dividing the data in such a way that it can be put back together. As in so much of contemporary wirelessness, military research has played a critical role here. Multicarrier modulation as developed in the 1950s could only be used for military communications because of the technical difficulty of implementing it. Dating from the mid-1960s, the FFT is widely used in signal analysis in many scientific and engineering settings.[8] Algorithm textbooks usually devote a chapter to the task of describing why practical applications of the FFT "demand the utmost speed" (Cormen and Cormen 1990). The FFT is a relatively quick way of computing the Fourier transform. Typically, for instance, as in "spectral analysis," the FFT displays the spectrum of component frequencies present in a signal as a series of bars on a bar graph–type display (see figure 3.3).

A visual display of this process can be seen in software music players such as Windows Media Player or VLC. They usually offer a visualization option called something like "Spectrum" that performs a Fourier analysis of the current media stream. The underlying mathematics of the Fourier transform has been known since the early nineteenth century: it can be written as a function that equates an arbitrary waveform with a series of periodic waves (sinusoids) of different frequencies and amplitudes. Some elements of the series of component waves make more important contributions to the overall shape of the original waveform than others. Added together again, these sine or cosine waves more or less exactly reconstitute the original signal. Crucially, a Fourier transform is usually understood to turn a waveform that varies over time into a set of component frequencies. The time axis on the right-hand side of figure 3.3 has been replaced by a

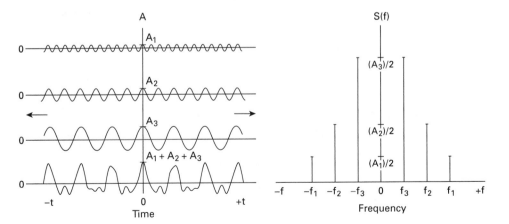

Figure 3.3
Fourier analysis

scale of different frequencies. The Fourier transform is a transform because it reversibly switches between "time" and "frequency" domains. Communication or audio engineers typically use Fourier transforms to analyze the "spectral density" of signals. They can show how energy is distributed in different frequency components of the signal. The FFT, dating from the mid-1960s (Cormen and Cormen 1990, 800), is an algorithm that allows the transform to be computed efficiently, and the availability of FFTs could be regarded as an indispensable component of signal processing of images, sound, and communication more generally. Unfortunately, any straightforward comparison between the role that FFT plays in digital signal processing of sound or images and wireless networks is marred by the backward use of the Fourier transform in wireless communication. Wireless signal processing uses the FFT back to front. That is, wireless signal processing purposely transforms a well-separated and discrete flow of information—the digital data stream—into a single complex signal. At the receiver, the FFT is then used to analyze the signal and divide it into discrete components that can be reordered as a digital data stream. In contrast to the standard and familiar uses of FFT in analyzing important components of a signal, in wireless networks, FFT is used to synthesize signals in such a way as to make them more complicated, more diffuse.

What would this loading of many tendencies onto a single wave achieve? Would it not increase the possibilities of confusion or interference? In the space-time of the transform, the digital ideal of discrete, linear flows of bits or bytes is suspended. The FFT as used in wireless DSP allows many different digital signals to be enclosed in a signal envelope and then extracted from it again. The FFT as used in wireless networks is like a courier company that only uses bicycle couriers in urban areas due to the density of traffic and the problems of parking. It distributes deliveries across many different small couriers or carriers rather than to moving large loads in a single movement. The FFT (and its many variants) as used in wireless networks exhibit a dispersed way of inhabiting the crowded, noise space of electromagnetic radiation. It splits the data stream into dozens of separate signals of slightly different frequency that together spread out across the available spectrum. This is done in such a way (via modulation techniques that I leave aside here) that many different transmitters can be transmitting at the same time, on the same frequency, without interfering with each other. Wireless transmitters are better at inhabiting a crowded signal spectrum when they do not try to keep themselves apart from each other, but actually take the presence of other transmitters into account.

The commercially important technique of orthogonal frequency division multiplexing (OFDM) used in 802.11g networks puts a data stream on approximately fifty component frequencies or "subcarriers." One of the benefits of spreading a single high-speed data stream across many signals ("wideband") is that each individual can carry data at a slower rate.[9] Because the data are split into fifty-two different signals, the symbol rate on each signal can be much slower (1/50). That means that the individual bits of data can be more widely spaced apart in time. The IFTT/FFT coupling, then, is used to slow down the time of wirelessness rather than speed it up. This spacing of data reduces one source of interference. In environments where they are many obstacles to signals, intersymbol interference (ISI) can occur. One symbol arrives at the same time as another because it comes from a reflection of the original signal. This is an especially bad problem in 802.11b networks, which use a different way of dividing up the signal (direct spread spectrum signal or DSSS). Both OFDM and DSSS allow many transmissions to use the same frequency, but they respond differently to the presence of echoes and reflections. IFFT is used to superimpose many signals onto one signal. That is, it takes the fifty or so different subcarriers, each of which has a single slightly different, but carefully chosen frequency, and combines them into one complex signal that has a wide spectrum. That signal fills the available spectrum quite evenly. In terms of spectrum, the waveform that results from the IFFT looks like "white noise": it has no remarkable or outstanding tendency whatsoever except to a receiver synchronized to exactly the right carrier frequency. At the receiver, this complex signal is transformed, using the FFT, back into a set of fifty separate data streams, which are then reconstituted into a single high-speed stream.

There is something quite counterintuitive in this transformation of a stream of digital data into a signal that looks like white noise, and the filtering of this noise back into a data stream. But in terms of radical empiricism, the presence of the FFT in wireless computation is significant. As digital data streams through the IFFT process and then across the air interface, it is heavily structured by microrelations that intensify conjunctive relations between transmitter and receiver. The pristine homogeneity of digital data stream mutates into a much more complicated topological structure fanning out across dozens of frequencies. The FFT shapes in certain ways what is directly experienced by expanding, rather than reducing, the relations between signals as they pass between transmitters and receivers. It constructs both transmitter and receiver so that they "neither admit into their constructions any element that is not directly experienced, nor exclude from them any element that is directly experienced" (James 1996a, 10).

When this now well-understood technique is brought into play, how does it affect a wireless signal? In what way does the FFT generate a conjunctive envelope? We have already seen that James describes transitions in experience using the metaphor of a "differential quotient." The differential quotient expresses the variations in rate and direction of a waveform. Such a conception of experience accommodates unexpected changes, shifts in direction, and altering circumstances as normal events. Its responsiveness to change is apt for networked, urban ecologies, where tangential encounters, veering off, and crowding in abounds. The cardinal consideration for a radical empiricist account of DSP is that the FFT literally treats all time-varying signals as expressing a set of tendencies. The relative size of component waveforms index the principal tendencies of the images, sounds, or other signals in process. In a sense, Fourier analysis and the Fourier transform express the waveform as *a set of partial tendencies* (partial differentials) rather than a single tendency. They literalize James's metaphor of experience as a waveform in a limited domain. The component waveforms are relatively simple to manipulate, to record, and to recreate because they are simple regular waveforms. Once a signal has been transformed into a set of components, these components can be assembled, circulated, and synthesized again in different ways. The FFT detaches the signal from its immersion in one flow of sensations, perceptions, gestures, and movements, and wraps it in another, that of a signal-processing environment.

James offers the differential-quotient-wave metaphor as a way of expressing "experiences of tendency." Tendencies and partial tendencies produce "variations in rate and direction" that are "sufficient to act upon" (p. 69). When James wrote of "a 'differential quotient' becoming 'conscious of itself'" (p. 69), he could well have been describing what the FFT brings to wireless signals. A waveform produced by FFT is "conscious of itself" to the extent that the data stream entering at the start of the process has been parceled in such as way as minimize the effects of their inevitable collisions with and alteration by other signals in a given setting. In the ways it spaces and times the signal, the algorithm imprints into the waveform a certain awareness of possible encounters as well as variations in direction and speed typical of urban life and mobilities. Any contemporary sense of media streaming or constantly-on communications in the city relies on such differential quotients and tendencies. These tendencies express relations of many different kinds: temporal processes, spatial gradients, and felt intensities. In proposing that James's notion of experience also encompasses a set of algorithmic processes, I am not arguing that radical empiricism provides a useful metaphor for understanding wirelessness. The claim here is stronger: wireless communications form an elementary component of contemporary experience. The felt reality of experience is interwoven, at least at the fringes of perception, with the conjunctively structured envelope of waveforms. Spatiotemporal conjunctive micromanifolds permeate contemporary experience. Only if we take that structuring of relations into account can we develop nonreductive accounts of wirelessness.

Ultraconjunctive Networks: The Viterbi Algorithm

How does a world hang together in the pure experience of radical empiricism? How do many tendencies occur at once without becoming chaotic sensation? James attributes the consistency or cohesion of a world to patterns of conjunctive relations. Conjunctive relations connect the world as "concatenated union." There is a characteristic flatness to many of the formulations of being-with in James's thought that conceals a sensitive awareness of a relational plurality and entangled pathways. In wireless communication, nearly all signals are marked by the presence of other signals. The situation is overwhelmingly relational in comparison to the relatively narrowly constricted flows of networks. This openness was known from early wireless experiments onward. The effect is especially great in urban zones, but also in deep space missions. Due to crowding of signals by many other forms of electromagnetic radiation, "severe channel" conditions often prevail in cities (and in deep space). In contemporary cities and built environments, the pressure exerted by the ideal figure of the network means that more messages have to move in more directions. With the growth of network environments in the last several decades, the problem of coexistence has taken on heightened urgency: How to transmit signals without destroying other people's transmission capabilities? The military-state solution has been to requisition large portions of the electromagnetic spectrum for exclusive use. Civil society has to work out how to share the spectrum. The limited spectrum made available by states to wireless networks needs to be habitable by many.

We could say that wireless signal processing solves this problem for networks by remaking the flow of data in an information network into very different images. It sets itself the task of discerning a path that passes from one point to another amid a myriad of crowded paths. It pays a high cost in technical complexity in order to hold the ideal of a network together with urban space. We can see this complexity in the case of the Viterbi algorithm. This algorithm occurs in many network media as well in bioinformatics. The algorithm dating from 1967 (Viterbi 1967) is widely used in telecommunications networks to maintain the sequence of information flows. It has become a fundamental element in commercial wireless, satellite, and space communications. Andrew Viterbi, a now retired telecommunications engineer, designed the algorithm and started a company (Qualcomm 2005) that designs and fabricates semiconductors based around the algorithm. In telecommunications applications, these chips enable satellite, cellular phone, and wireless networks to communicate despite high levels of electromagnetic noise. In bioinformatics, the Viterbi algorithm is currently used to find weak similarities and evolutionary kinship between protein or amino acid sequences within families of functionally related biomolecules (Lesk 2002).

If the FFT is like a courier company that resorts to bicycle couriers in downtown areas, the Viterbi algorithm is like the recipient of a deluge of partly addressed parcels trying to work out who they are meant for, and which was sent first. The Viterbi algorithm deals with situations where the source of signals cannot be directly observed. This could be because a signal transmitted in the crowded environment of a city (or interplanetary space) has reflected or bounced off some obstacle, and this changed the order of information received at the receiver. The Viterbi decoder has the task of rediscovering the sequence as transmitted. It was originally conceived as an error-correction scheme for noisy digital communication links, and then found application in digital cellular telephones, cable modems, satellite and deep-space communications, and 802.11 wireless LANs.[10] In general, the algorithm finds the most likely series of "hidden states" that could have given rise to the observed events.

The basis of the Viterbi application to language, life, and media hinges on the idea of finding the most probable *hidden states* that would account for the currently observed behavior in a system. For instance, in an 802.11 wireless network (IEEE 1999), in a GSM cellular telephone network, or in any situation where severe channel conditions prevail, the data itself may have changed during transmission. A short burst of interference may introduce errors in the data stream. To cope with this, all data are encoded using "convolutional coding," a type of "forward error correction." Convolutional coding can be understood as a kind of sequence label imprinted within the signal itself rather than on its packaging. The IEEE standards

document for Wi-Fi networks enjoins engineers strictly: "The DATA field
. . . shall be coded with a convolutional encoder of coding rate R = 1/2,
2/3, or 3/4, corresponding to the desired data rate. The convolutional
encoder shall use the industry-standard generator polynomials, g_0 = 133
and g_1 = 171, of rate R = ½" (IEEE 1999, 16).

In convolutional coding, the computational processing capacity of the
transmitter and receiver is used to compensate for noise and errors produced in the transmission channel. When encoding the information, the
transmitter infuses extra data in the sequence of data by applying a carefully chosen mathematical function: the "generator polynomial."

Convolutional coders take their name from the way they base what they
transmit at the current point in time on what has been transmitted earlier.
They begin to build a "state of mind" concerning what has preceded the
current moment in the data stream: "We have states of mind and so do
encoders. We are depressed one day and perhaps happy the next from the
many different states we can be in. . . . We can say that encoders act this
way too. What they output depends on . . . their state of mind" (Langton
1999, 3).

The "convolution" consists in this folding of the data stream to
incorporate information about what was transmitted before. Each packet,

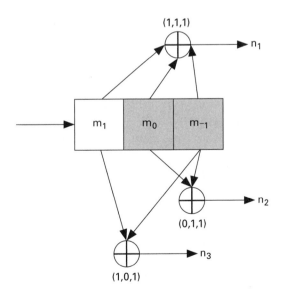

Figure 3.4
Convolutional encoder ("Convolutional Encoder Diagram" 2007)

datagram, or frame represents not just information, but a relation to what came before. A convolutional code enlarges the frame of the now itself in specific ways. Intensified movements in transmitter and receiver compensate for disturbances or errors arising from what cannot be made fully part of the system: the pathway along which signals propagate. Convolutional codes assume that communication comprises a process of conjunction that cannot be fully controlled or determined and that therefore cannot be definitively transmitted. Acceptance of this state of affairs begins to undermine a mechanistic understanding of algorithms as repetition. The Viterbi algorithm used in decoding a wireless signal couples with convolutional codes that complicate the forms of repetition in the data stream. Together, convolutional coding and Viterbi decoding allow wireless computation to embrace the *stochastic* processes occurring in the "severe channel" conditions. In stochastic processes, the next state of the system is not fully determined by the previous state. A snakes-and-ladders game (also known as chutes and ladders) is a stochastic process because the next move is partially determined by a dice roll. The stochastic character of the Viterbi algorithm both determines the qualitative effect of repetition and alters the terrain on which machine time moves.[11]

We can begin to see how this twist might be useful in communication systems where there are many obstacles to transmission and reception. In telecommunications, including satellite, cellular, and wireless network communications, many different kinds of interference and noise affect the propagation of digital signals. The task is to work out the most probable sequence of signals that could have given rise to the observed signals. Again, counter to the images of strict determinism sometimes associated with digital technologies or information systems, the algorithm engages with an unpredictable and intrinsically dynamic environment. It moves to the fringes of mechanical action, where repetition and difference begin to blur.[12]

Put in James's terms, the Viterbi algorithm in combination with convolutional coding allows wireless computation to support many possible conjunctive relations between successive parts of a signal. Even if not all parts of a signal are experienced due to "severe channel conditions," the paths or conjunctive relations that connect parts can be used to keep the world hanging together. The encoding-decoding algorithm expects lots of errors, but builds ways of fixing the errors into the data stream. Rather than trying to preserve every signal in its own place as a thing in the world (disjunctively), it algorithmically fosters *conjunctive relations*. We could say that in the Viterbi algorithm, "Relation is immediately perceived *as such.*

A relation is not a secondary production of association" (Massumi 2002, 230). Convolutional coding creates a great many relations internal to the signal, so that their relationality exceeds any possible interruption by another signal.

"Manhattan Distance": Multiply Relationality in Order to Diminish It?

Manhattan distance. Definition: The distance between two points measured along axes at right angles. In a plane with p1 at (x1, y1) and p2 at (x2, y2), it is $|x1 - x2| + |y1 - y2|$. (Black 2006)

The Viterbi forward error-correction process injects so many relations into the data stream that it creates new logistical problems: How to manage the proliferation of conjunctive relations—the convolutions—in the data stream. There is a major obstacle here. James's radical empiricism pivots on the idea that no one path of conjunctive relations running between things will have more reality than any other. He writes that "*some* path of conjunctive translation by which to pass from one of its parts to another may always be discernible" (p. 107). But don't many different possible paths run between two parts of a signal, just as many different paths run between the squares on a snakes-and-ladders board? To decide how two things relate to each other, a question must be answered: Which path is most discernible?

James's "roughly ascending order of intimacy and inclusiveness" of types of conjunctive relations comes into play: "with, near, next, like, from, toward, against, because, for, through, my" (p. 45). These words designate types of conjunctive relations arranged in order of increasing intimacy. The Viterbi algorithm can be seen as a way of sorting relations between bits in a data stream in order of increasing intimacy. Although bits inevitably arrive in a sequence and hence have conjunctive relations of "withness" or "nearness," the algorithm discerns the relation between bits that produces the greatest "intimacy and inclusiveness" in the stream. This is taken to represent the most likely version of the signal transmitted.

Importantly, this whole process of encoding and decoding imagines the data stream as an itinerary of movements in a city. In classifications of algorithms, the Viterbi algorithm that decodes the convolutional-coded signal is broadly regarded as a "dynamic programming" algorithm. The Viterbi algorithm descends from the field of operations research, a field originating in the logistics problems of the Second World War. Computer science textbooks usually contain at least one chapter on dynamic programming since it is now a classic algorithmic technique. Dynamic programming addresses the problem of finding optimum routes or paths through networks or grids. Logistical problems concerning networks, ranging from flight scheduling to text searching, make use of dynamic programming approaches. Dynamic programming models all problems in terms of traversals of directed acyclic graphs (DAGs), or networks in which movement can never go in circuits. By transforming problems such as sequence comparison into a graph/network data structure, and carefully planning the order in which computations are carried out, dynamic programming algorithms drastically reduce the time needed to find the optimal itinerary. The typical contemporary metaphor for explaining dynamic programming is the problem of moving through a city grid. Typically it asks how a taxi or a tourist in Manhattan could visit the most attractions with the least driving or walking. Given the gridlike street layout of Manhattan, there are many different paths that could be taken to include MoMA, the Empire State Building, Times Square, and Wall Street. A slightly older metaphor asks how a traveling salesperson can visit all the towns in the region doing the least driving.

Many different situations can be translated into the model of Manhattan's streets, even nonspatial ones. The basic idea is that a sequence of information can be represented as a path on a grid. An algorithm textbook for bioinformatics claims that "development of new sequence comparison algorithms often amounts to building an appropriate 'Manhattan'" (Jones and Pevzner 2004, 160), emphasizing the imagined connection between urban space understood as a data structure and algorithms understood as itineraries or paths. The movement of tourists visiting attractions dotted on the grid of Manhattan's streets provides a useful spatial model or "data structure" for other forms of movement, so much so that "Manhattan distance" is a standard way of referring to this kind of movement. The Viterbi algorithm maps the many possible ways of arranging the information it receives onto a grid or "trellis" (see figure 3.5).

Figure 3.5 offers a very rough idea of how Viterbi decoding processes conjunctive relations: if a received sequence of bits does not fit this graph, then it was received with errors, and we must choose the nearest *correct* sequence of bits that fits a path in the diagram. Actual decoding algorithms exploit this idea. The dynamic programming technique has been imitated and adapted in applications across bioinformatics, logistics, chess, and telecommunications because it lays down an especially optimal concatenation of steps between source and destination. This concatenation does not rely on a brute-force technologically determined acceleration in hardware

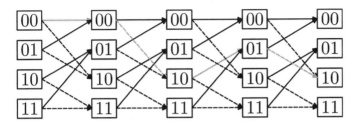

Figure 3.5
Conjunctive relations in Viterbi decoding ("Viterbi Algorithm" 2008)

or processor speed. Rather it relies on bringing the problem domain—how to move through a city while keeping connections—into the solution. In terms of the city-itinerary metaphor, it registers the fact that in the many different possible itineraries, most will follow partly in the footsteps of others. Rather, by virtue of a reversal that lies at the heart of dynamic programming algorithms, the Viterbi algorithm traverses the interval between input and output counterintuitively. That is, it carries out calculations in a different order from that dictated by a practical sense of the world as something to be acted on effectively. This limited part of the world undergoes topological transformations and reordering in time. The transformation of computational time wrought by dynamic programming proceeds via a process of intensification of movement that needs to be explained rather than simply attributed to algorithms or to abstraction per se.

The "hack" at the center of the "creative production of abstraction" (Wark 2004, 71) in the Viterbi algorithm is not based on a better representation of something—for instance, of the prototypical situation modeled in software. Nor is it based on a solution that puts an end to movement. Instead it brings the problem of how to move through a city deep into the structures of the conjunctive envelope. In finding the shortest path through a grid (the Manhattan distance) via several selected points, the problem is to minimize repeat visits. In this sense, the algorithm seeks to reduce repetition, not to multiply it. Dynamic programming is a tactic to avoid doing calculations that have already been performed or to avoid returning to points in the path that have already been visited. Every repeated calculation equates to the pedestrian backtracking over the same ground. To minimize repetition, dynamic programming abstracts from the particular to the general. Rather than trying to solve the specific problem of how to get from one point to another most directly (for example, the minimum number of blocks to walk to visit all the popular sights), dynamic programming algorithms build a table or array containing the best scores for all possible movements (for instance, between all known tourist destinations in Manhattan). A new, conventional order, that of the table or matrix, replaces the topology of the network or the topography of the city. Movements in the ordered space can be carried out without repetition because the table precalculates the cost of different movements. By solving the general problem of movements between many different points, dynamic programming makes finding the best movement between two chosen points much more efficient. By envisaging all possible moves in advance, it accelerates the process of finding which one in particular is the shortest. The creative "hack" is to reverse the intuitive approach that would start with the particular and move to the general case. Dynamic programming imagines the general case of movement in order to discern a particular movement that maximizes the conjunctive connections. This description of the algorithm, it should be remembered, renders the problem of what sequence of bits have been transmitted as what is the shortest path passing a set of destinations in a city. To communicate in the city, a model of urban planning and delivery logistics has been internalized in the algorithms of signal processing.

Being-with Is Multipath?

I have been examining some of the transitions occurring in wireless digital signal processing and in the billions of wireless chips. These processes can be experienced directly, although the feelings associated with them remain peripheral most of the time. These feelings and sensations can become the material for diverse aesthetic experiments (for instance, see the works at the "Spectral Ecology" event in Riga, Latvia, in 2007 (e.ngo.org 2007)). But any such experiments must contend with the fact, as we saw in the case of FFT, that the conjunctive envelope of wireless DSP has been constructed as a form of consensual "noise." It supports consensual copresence without conscious association.

The two algorithmic processes discussed here express tendencies that define the conjunctive envelope of wirelessness: waveforms function as a way of dealing with variations in rate and direction; concatenated paths make worlds, or parts thereof, hang together. In light of these tendencies, we are now in a better position to comprehend why the PicoChip or any other contemporary wireless chipset might need so many processors, or why chipsets multiply in wireless networks. Wireless signal processing

rewires the interstices, interiors, corridors, and avenues of built environments as densely conjunctive spaces, by treating them as spaces populated
both by tendencies to change and by proliferating pathways. The abundant
calculation that DSP wraps around the network flows of voices, images,
data, or control seeks to fabricate a signal that can diffract through the
highly cluttered relational spaces of contemporary urban life. We have seen
that the algorithmic architecture of wireless signal processing envisages
constant change at several levels: in rate and direction of movement, in
the presence of others, in the trajectory of the signal at any given moment.

At their core, I would argue, algorithms in concert with chipset architecture perceive contemporary cities as proliferating movements, as full of
encounters with obstacles and blockages, and as replete with connections
between variously intimate and strange others. Each component of the
conjunctive envelope addresses a different aspect of this perception. For
instance, the problem the Viterbi algorithm solves is one of sequence:
given that the order of events can easily be affected by different settings,
which is the best ordering of events? In James's terms, the algorithm discerns a pathway of concatenated union of greatest inclusiveness. Because
they increasingly construct perceptions or images of the city within the
signal envelope, we could say that the algorithms generate microworlds.
These microworlds possess qualities that affect any contemporary experience of force, movement, feeling, and duration. Importantly, the conjunctive envelope synthesized in wireless signal processing generates an
expectation of "more to come." The conjunctive envelope accommodates
and opens up further transitions, new conjunctions between places, events,
and movement, and then recursively generates much more communication about those conjunctions. More things become enveloped by intermediate paths and air interfaces. Like urban life in general, wirelessness is
an experience of something more to come, of what James (1996a, 237)
describes as the conjunctive feeling: "While we live in such conjunctions
our state is one of *transition* in the most literal sense. We are expectant of
a 'more' to come, and before the more *has* come, the transition, nevertheless, is directed *towards* it."

Although mostly not consciously experienced, the conjunctive envelope in Wi-Fi and other wireless standards intensively reorganizes spatiotemporal processes and tendencies in ways that are not reducible to the
comparatively static figure of the network and its flows. The conjunctive
envelope generated in wireless signal processing exceeds networked
communication. It can be seen as extraneous to the network to the extent
that it exposes forms of being-with that cannot be contained in network

structures of nodes and links. Wireless signal processing embraces tendencies toward excess belonging-together. Almost contentless and transparent, conjunctive relations such as "with," "near," "between," "and," "beside," and "next to" have critical importance in the everyday fabric of connectivity. However, these relations are difficult to theorize without falling back into psychological or sociological accounts of experience. In attempting to resist that, this chapter has untangled two significant algorithmic processes enveloping conjunctive relations within wireless networks. It has suggested that the convoluted character of these algorithmic processes literally proliferates conjunctive relations, events, and phenomena in order to animate wirelessness with variations that multiply and diversify. We could see these bare conjunctions as ontologically vital signs of contemporary being-with. If as Jean-Luc Nancy (2000, 3) argues, "Being cannot *be* anything but being-with-another, circulating in the *with* and as the *with* of this singularly plural coexistence," then wirelessness exposes in intensified forms the circulation of "with." The next chapter examines how even the bare exteriority of *with* supports very different forms of circulation.

4 Devices and Their Boundaries: Inventing Wireless as "Vast Space"

For most people, wirelessness means devices—mobile phones, routers, game consoles, media players made by companies such as Apple, D-Link, Motorola, Belkin, Samsung, Netgear, Linksys, or Dell. However, abstract processes—an enveloping conjunction of relations coalescing around problems of spacing, departure, arrival, proximity and being-with others—attach to all contemporary wireless devices. Wireless devices, enmeshed in the psycho-infrastructural economy of network media , undergo constant transitions, processes of modification, and experimentation. Wirelessness agitates, frustrates, and bores, as well as satisfies, excites, and gratifies. These feelings often surface around devices. While changeability is not unique to wirelessness (since it can just as well occur in relation to climate change, reality TV, computer games, finance markets, art exhibitions, political life, biomedical research, music, conferences, etc.), devices present tangible traits that allow wirelessness to be investigated in the making.

A radical empiricist understanding of wirelessness as "turbid sensation" of change would pay attention to how devices are made and unmade in experimental—and antiexperimental—practices.[1] A great diversity of practices could be discussed here. They range from the ways people mundanely modify, configure, or substitute settings on wireless or mobile gadgets, to the estimated 250 million shānzhàijī or "bandit phones" made in small workshops in southeast China during 2008. In the middle ground, we could discuss the post-dot-com bubble valorization of "making" and DIY technology in North America, Europe, and Australia. The online and print magazine *Make: Technology on Your Time* (O'Reilly Media Inc. 2007) enthusiastically endorses modifying, altering, and reconfiguring (mainly) digital hardware and software in low-cost projects.[2] The advent of open wireless platforms such as the *Open Handset Alliance* and Google's Android for mobile phones (Open Handset 2009) further exemplifies the shifting boundaries of wireless devices. Although all the chapters in this book could be understood as concerned with experience in the making, this chapter is directly focused on practical work done on particular Wi-Fi devices such as wireless routers. In contrast to the largely invisible intricacies of the billions of mass-produced semiconductor chipsets and algorithmic calculations of the previous chapter, devices such as routers modulate wirelessness into many tangible and intangible forms—products and services, gadgets, and infrastructures. The boundaries, physicality, networkability, and individuality of wireless devices undergo extensive experimentation. Viewed as modulations of the conjunctive envelope in network settings, wireless devices attract extensive efforts at configuration, modification, experiment, and commodification. These efforts are not trivial or superficial. They are often highly generative, deep-ranging, and potent, and quickly trickle back into commercial products. The chapter, then, takes a particular interest in the ways that people handle wireless devices and equipment, and what happens when they do. The materials presented in this chapter focus on a class of devices known as wireless routers, and the Linksys WRT54G wireless router in particular as it appears in policy documents, wireless how-to books, and software such as *OpenWRT* (OpenWRT 2008). (A similar analysis could focus on mobile phone handsets or other wireless devices.)

The Wireless Router as Progression from Bad Network to Good Network

To progress in knowledge . . . signifies to pass from a bad network to a good network. (Latour 2007, 28)[3]

Much of the work done on wireless devices focuses on the gap between what a network is and what it is not, between what we might call a "bad network" and a "good network." The difference between a bad network and a good network drives many interesting dynamics of wirelessness. Take, for instance, the popular WRT54G wireless routers made by Linksys, a brand name now owned by Cisco Networks, a globally significant manufacturer of networking equipment. A router is a piece of network equipment that decides where on the network to send packets of information it receives. (More technically, a router connects two networks or "subnets." It might connect two wireless subnets or, more typically, a wireless subnet and a wired subnet.) Until the arrival of wireless devices, homes and cafés did not need routers. Their arrival suggests that networks, in the plural, have entered domestic spaces. It might be hard to feel at all excited by Linksys routers since they do not seem to do very much. Their chunky blue-and-black plastic enclosures squat on shelves and tables in houses, apartments, cafés, and offices. As devices, they entirely lack the sharp design edge of the latest Motorola Razr mobile or Apple iPhone. Their half-dozen small flashing LEDs hardly match the media extravaganza on offer in a Sky+ digital video broadcast satellite television box. They do not upend the living-room furniture like a Nintendo Wii game controller (itself a Bluetooth wireless device). Yet Linksys routinely sells several hundred thousand routers each month (Asadoorian and Pesce 2007, 3), so they can be found in the millions. Given its innocuous status, you would think that Linksys could control the production of the Linksys WRT54G wireless router in such a way as to guarantee a predictable modulation of wireless signals and calculable growth of the Internet. Actually, the Linksys WRT54G wireless router encompasses a series of devices built of many components that are constantly changing, and not always according to the manufacturer's wishes. Since 2002, there have been approximately fourteen major versions of the WRT54G, and each has a number of minor versions, so that overall there are around forty-five versions of this one product line, the Linksys WRT54G (the *Wikipedia* entry gives a good sense of the versioning; see "Linksys WRT54G Series" 2008). The series of devices are all called WRT54G (**W**ireless **R**outer **54**M/bs 802.11**g**). The devices all look broadly similar, and ostensibly, they do similar things. It is actually sometimes quite difficult to know which one you are buying.

As a variety of pragmatism, we would expect radical empiricism to say a lot about practical differences. Pragmatism has primarily been understood as a method of settling metaphysical disputes in terms of practical differences. As James (1978, 16) define it, "The pragmatic method . . . is to try to interpret each [metaphysical] notion by tracing its respective practical consequences. What difference would it practically make to anyone if this notion rather than that notion were true? If no practical difference whatever can be traced, then the alternatives mean practically the same thing, and all dispute is idle."

For *notion* in the above formulation, we might substitute *device*. The pragmatic method would ask: What difference does it make it to anyone if one version of a wireless router rather than another is made, sold, bought, or brought into existence? Any such difference would only occur if people had different ideas of what makes a bad network and what makes a good network. If they want to move from bad to good networks, a device might make the difference. A gap exists between the ideal of a totally networked world and a provisional, unstable reality tangled with wires, buildings, everyday habits, and the presence of others, between, in short, meaning and *praxis*. That gap exists, according to Jean-Luc Nancy (2007,

54), because meaning does not exist outside practice: "Meaning is always in *praxis,* although no practice is limited to enacting a theory and although no theory is able to diminish practice." Things and practices cannot be fully interpreted or rendered meaningful, and conversely, meaning cannot be entirely put into practice. Consequently, an excess immanent to experience generates work "whose principle is not determined by a goal of mastery (domination, usefulness, appropriation), but exceeds all submission to an end—that is, also exposes itself to remain without end" (Nancy 2007, 54). Although mastery, usefulness, and "appropriation" (of value) are writ large on wireless devices, some of the work done on them cannot be captured by such ends.

From the outside, almost everything about the router is generic. For example, some lights flash to indicate power, connectivity, and network activity (see figure 4.1). Apart from some general physical similarities (although even these are subject to radical alteration, as we will see), the only thing that links different versions of the WRT54G is the quasi-abstract center of envelopment described in the last chapter—the IEEE 802.11a/b/g (and n) WLAN standard. In different ways, all WRT45Gs instantiate that convoluted center of envelopment. Various features of the objects refer to the signal processing and networking protocols. Most visibly, the name of the device, and various Wi-Fi Alliance–approved labels on the different versions all refer to "g" or specifically to "802.11g," a somewhat later, higher-speed version of 802.11 wireless networks.

Despite being so generic, or, perhaps because they are so generic, these devices harbor surprising degrees of relationality and support a wide variety of experimental (and antiexperimental) modifications. Unexpectedly sophisticated webs of relationality unfold out of and around them. They are constantly being remade, both by Linksys and its component suppliers, and by others in, for instance, various antenna modifications, in changes to the firmware, or in replacing the enclosure. Practices of making and unmaking (preventing change) deeply affect how the contemporary experience of wirelessness takes place.

Isn't experience something that is just pragmatically given, that we perceive in some ways, that we ideate or remember in others? And in relation to wireless communication, doesn't the presence of the wireless networks in houses, cafés, airports, hotels, and city streets simply deliver connection to information networks in a way that we can quickly take for granted, or that we at least hope to take for granted? There are number of problems and obstacles in taking experience as just given. First, wirelessness has a texture that intensifies certain aspects of experience. Second, in

Figure 4.1
A Linksys WRT54G wireless router.
Credit: Jonathan Zander.

contrast to phenomenology and the many sociological approaches that draw on it to anchor experience in the living, speaking subject (Derrida 1973), the impersonal and material aspects of wirelessness fold in from the edges of experience. They are not given in experience, nor outside it. While pragmatic thought in general is often taken as a method for testing ideas, it harbors within it, and especially in James's version of it, a much more animated, effervescent account of experience. James's characterization of the weaving together of transitions in experience is symptomatic of this: He writes: "According to my view, experience as a whole is a process in time, whereby innumerable particular terms lapse and are superseded by others that follow upon them by transitions which, whether disjunctive or conjunctive in content, are themselves experiences, and must in general be accounted at least as real as the terms which they relate" (James 1996a, 62).

Here, the emphasis lies on the need to count transitions as real as any content of experience. James affirms the "reality of change" (Massumi 2002, 201). His radical empiricism makes a transition into a leitmotif of experience. If wirelessness is typical of contemporary experience, then it cannot be understood without reference to the vectors and gradients of change that it enfolds and compresses. We can also see radical empiricism, as this chapter does, as a name for any exploration of the boundaries, contours, neighborhoods, and limit points of making and remaking things in transition. The "innumerable particular terms" of wirelessness include many specific practices as well as pieces of hardware and software caught up in "transitions" that cut across and recombine things and feelings, gestures and processes.

Who Sets Boundaries on Devices?

Wireless spaces fluctuate readily, sometimes even with the weather, but particularly in the presence of others. Changes affect wireless devices, even just a single wireless device such as a wireless card (see figure 4.2), differently.

There is an uncertain spatiality to wirelessness that comes from the encounter between the conjunctive envelope described in the previous chapter and network topologies. In network topologies, connections and relations are often centrally managed. The uncertain limits of wirelessness have already been a matter of public concern in various forums. For instance, a "spectrum-commons" debate began in the late 1990s, particularly around the auctions for the mobile phone spectrum. This debate was

Figure 4.2
"AirStation" in the kitchen.

an adjunct to other wide-ranging debates about intellectual property and information networks that were occurring at that time. In 2003, a well-known technology analyst, Kevin Werbach, published a report titled *Radio Revolution* (Werbach 2003) arguing that existing regulation and allocation of the radio-frequency spectrum was based on a misunderstanding of wireless communication. The report entered into the wider spectrum-commons debate by arguing for changes in the way the U.S. Federal Communications Commission (FCC) allocates frequencies. It was structured around a series of analogies and explanations of how wireless technologies have changed. According to Werbach (2003, 2), "Our intuitions about wireless, by and large, are mistaken. They are based on outdated technologies and inaccurate analogies. If we hope to move forward in exploitation of the airwaves, we must take a step back. We must understand wireless communication for what it really is. And then we must re-evaluate our assumptions about what it could be."

Werbach argues that existing spectrum allocation and licensing regimes mistakenly focus on an abstract territorial concept of spectrum (as a limited, even scarce resource that needs to be parceled out carefully in order to avoid conflict, competition, and interference that would render it useless to everyone). Instead, according to Werbach, in order to "move forward in exploitation of the airwaves" we need to embrace a "new dynamic paradigm" in which "more than one service can occupy the 'same' spectrum, in the same place, at the same time. The frequencies that now carry one signal could someday carry thousands . . . or billions" (p. 3). The new dynamic paradigm focuses on the behavior of devices: the "key question in a world of dynamic wireless systems is how to set the proper boundaries on how devices can operate" (p. 16). The ins and outs of the spectrum-commons debate need not greatly concern us here. That debate, and the attempts to develop new paradigms of wireless communication within it, suggest that wirelessness as a contemporary mode of experience concerns a topologically problematic space.

Werbach argues that rather than asking how to allocate radio spectrum, the key question should be "how to set the proper boundaries on how devices can operate." Although Werbach aims to reform U.S. regulatory practices, the question of the "proper boundaries" on wireless technology is much wider than a regulatory issue. There are many ways of setting the boundaries on how devices operate. And there are many different people or groups of people who could be involved in setting or changing boundaries. Many of these ways of setting boundaries are part of the devices themselves. Boundaries are not simply imposed from the outside. They can be altered by modifying hardware (antennae, enclosures, etc.), changing or upgrading software (the "firmware" that the devices use), configuring the many security and access settings on the devices, trying different locations for the devices (on roofs, on street poles, in cars, etc.), as well as by federating devices in new sets and groupings (as did the many community and wireless mesh networks), or by building wireless devices from standard components (as do the shānzhàijī device makers in China). In a sense, wirelessness transpires as an experiment that explores different ways of altering the boundaries on how devices operate. This experiment also involves the experimenters. The question of who can alter the boundaries on devices is contested.

Wirelessness understood as an experiment in who today can alter what device boundaries diverges from a now more familiar interpretation of the relation between experience and modern infrastructure. It is widely accepted that mundane experiences of place, movement, and mobility rely on infrastructures that remain more or less invisible (but for a critique of this, see Edwards 2003). For instance, the social scientists Paul Dourish and

Genevieve Bell (2007, 417) argue, in the context of pervasive computing, that infrastructure shapes the fabric of experience:

By "the infrastructure of experience," we want to draw attention to the ways in which, in turn, the embedding of a range of infrastructures into everyday space shapes our experience of that space and provides a framework through which our encounters with space take on meaning. The experiential reading of infrastructure, then, sees infrastructure and everyday life as coextensive; accordingly, it encompasses not just technological but also the social and the cultural structures of experience in pervasive-computing settings.

In this view, infrastructure has both technological and sociocultural aspects. As technology, infrastructure shapes experience and materially supports meaning, itself deposited in social and cultural structures. While this makes sense and would be hard to disagree with, it still departs from two separate realities—technology that is more or less physical and experience that is more or less social and cultural.

Rather than saying that infrastructures such as wireless networks shape an experience of space, we could say that through wireless devices certain tendencies or potentials in contemporary experience are explored and spatialized. These tendencies or potentials are not always consistent with each other. From a radical empiricist perspective, infrastructure can readily invert into experience. Thing and thought are not in principle separable in radical empiricist understandings of experience. Precisely because they weave through experience, wireless devices are sometimes infrastructures and sometimes highly intimate possessions. The same device can be both. For instance, a small device such as the Novatel MiFi acts as a "personal hotspot" (Pogue 2009), and is both infrastructural and personal. Via wireless devices, what Dourish and Bell call the "embedding" of infrastructure into everyday space, I would argue, enters an unstable trajectory. Wireless equipment lies somewhere on the boundary between the proper infrastructures of modernity (roads, airports and railways, telephone exchanges, lines, transmissions towers and satellites, electricity pylons, water mains and gas pipes, etc.) and consumer electronics. Today, many wireless devices, especially relatively bland products such as access points like the Linksys WRT54G or the many gadgets with Wi-Fi built in, take the shape of small boxes with short antennae (although as the following chapters will argue, the shape of wirelessness is constantly reimagined in different forms— maps, networks, gadgets, etc.). These boxes look somewhat different from most other consumer electronics. Their relatively simple controls and small size mean that they can quickly fade into the background in most settings. In commercial and institutional settings, they can be found mounted high on the wall, or inside another enclosure (for instance, in the United Kingdom, the commercial wireless access points in thousands of bars and pubs lodge inside a gambling machine called "Who Wants to be a Millionaire" based on a popular TV show). There are many built spaces now populated by such devices in ways that we are scarcely aware of (Thrift 2004), and this trend shows no signs of abating as femtocells and picocells promoted by telephone companies arrive on the doorstep to compete with Wi-Fi-based broadband. A phenomenology of wireless boxes mounted via wall brackets might conclude that wirelessness has a quotidian invisibility. It is practically bracketed out (although this bracketing out is constantly interrupted, as will be discussed in a later chapter). Sometimes, one way or another, we are compelled to perceive these boxes and make use of them in ways that shift quite dynamically. There is a truism in the phenomenology of technology and probably in the cultural and social studies of technology more generally that we become aware of technology only when it breaks down.[4] In fact, in relation to wirelessness and many other technical situations, a flickering oscillation between breaking-down, becoming-aware, and background-forgetting is more common. Organized forms of breakdown—hacking, copying, modification—attract energy, investment, and attention.

The devices that comprise the infrastructure keep changing, and this process of change is difficult to control. New components are constantly being added, and sometimes old ones are taken away by the manufacturers of consumer electronics. While some of this dynamism is due to the market conditions in which wireless equipment is built using standard mass-produced components, other parts of it are due to processes that overflow the market competition, and in fact subvert markets and branding. Here the Linksys routers are typical. They, like most consumer electronic devices today, are effectively scaled down computers, with their own memory, CPU, interfaces, and software. These components come together in the router from different sources. Sometimes the substitution of components flows in the wake of relatively small changes in the internal design of the object as engineers find new ways to cut costs, simplify fabrication, or enhance the behavior of the object. For instance, after the networking equipment corporation Cisco Systems, Inc. purchased Linksys in 2003, models of the WRT54G have fewer lights on the front but an extra button for "Secure Easy Setup." Viewed from the outside as a box, the router looks generic and static, and indeed has been mass-produced. Seen over time, however, the router has undergone a surprising degree of modulation, activity, and expression related to software cultures more generally. Sometimes new components come and open the device to whole worlds or realities that were previously not part of the object. (For instance, as discussed below, a larger flash memory might change the kind of software that can be installed in the router.) These changes alter the boundaries within which the devices operate.

If it seems that infrastructures exist relatively inertly in comparison to experience, we need only think of the work of maintenance and repair that goes on in contemporary cities around communication and transport. Stephen Graham and Nigel Thrift (2007, 8) argue that maintenance and repair form an increasingly important component of the life of "cities of repair" today. Frequent cycles of replacement and disposal of electronic equipment such as computers and telephones bring severe problems of configuration, maintenance, and upgrading, as well as geopolitical problems of waste management, energy supply, and pollution into play. This means that an experience of infrastructure includes transitions between working and not-working, between emerging and the taken-for-granted, between old and new: "The inherent and continuous unreliabilities within all infrastructure systems, which necessitate continuous efforts of repair and maintenance to actually allow them to sustain the distantiated connections and flows that they are designed to deliver, still tend to be rendered invisible both culturally and analytically" (Graham and Thrift 2007, 10).

While Graham and Thrift see the inherent unreliabilities of infrastructure as something largely negative, and as therefore tending to be "invisible both culturally and analytically," I would argue that in the certain respects wireless networks overflow this observation. Their unreliable transitioning has been heavily mediatized (see the following chapters), partly because their value and status as infrastructures have not been able to stabilize. For instance, in the case of municipal or metropolitan wireless networks, their outlines or boundaries as public, private, or commercial infrastructures cannot be readily fixed and have repeatedly shifted.

Occasionally devices affect infrastructure itself, especially if that infrastructure is "interwoven with the existing physical structure of space." We need not assume the possibility of concretely separating technology and experience at the outset. Experience, following James's account of it at least, cannot be easily corralled or concretized as either social or physical. Importantly, as discussed previously, experience has many impersonal aspects that do not feed directly into meaning, that hover on the edge of intelligibility as sensations or feelings. No doubt, social and cultural structures identify, fix, isolate, and abstract certain aspects of experience. But in terms of James's radical empiricist account of experience, the flow of experience relies on relations that themselves do not always appear as images, statements, or signs. These relations have vital importance in constituting the flow of experience. James (1996a, 48) writes that "continuous transition is one sort of a conjunctive relation; and to be a radical empiricist means to hold fast to this conjunctive relation of all others." The key argument here, then, is that in work done on wireless devices, experience, and the infrastructural conditions of experience intermittently and provisionally coalesce. The assumption that an infrastructure frames a single space comes into question around wireless devices, some of which intersect with multiple infrastructures on different scales (Bluetooth, Wi-Fi, GSM, etc.), and some link multiple infrastructures together.

Internal Breakdown: The Kludge

Wireless devices display equivocal status as infrastructure. They subject various infrastructures to variation. Network infrastructure in particular goes into transition in wirelessness. There are several dimensions to this transition at work in IEEE 802.x networks such as Wi-Fi and WiMax. This is not only because the repair, improvement, upgrade, and maintenance work constantly alters things. It is because the thing itself, the wireless device, was always an unstable composite charged by different, even contrary, tendencies. In 2003, a participant in the *Consume* project based in Greenwich, London, touched on this when he exclaimed during an interview: "802.11b is a kludge!" A "kludge," according to the *New Hackers Dictionary*, is a hacker term for "an ill-assorted collection of poorly matching parts, forming a distressing whole" (Raymond 1996, 221). At that time, there was so much to be seen and heard about the promise of Wi-Fi, and about how adaptable, powerful, and effective it is, that it was striking to hear Wi-Fi—a very highly promoted and arguably successful networking technology built into millions of systems—derided as a "distressing whole." Given that the interviewee was heavily involved in a relatively prominent, well-known community wireless project that had many operational nodes deployed and working, why should he call Wi-Fi a kludge?

Kludges do not always arise from a technical design defect. A long history of kludges constitutes the material culture of media. The quasi-geometrico-optical metaphor of "convergence" that is often used to describe contemporary media neatens up the unstable mixtures of audiovisual, communications, and computing techniques occurring for the past few decades as various sound, visual, televisual, textual, graphic, telephonic, and now radio media are folded into computing, and hence into other mobile and consumer electronics devices. They sometimes attest to the divergent realities, needs, and desires articulated together in technical objects. Whereas the notion of convergence emphasizes reduction to a well-defined context or state of affairs, "kludge" gestures toward *relationality*, to ongoing changes in nature stemming from juxtapositions. Massumi's (2000, 191) concept of relationality captures certain aspects of a kludge well: "Call the openness of an interaction to being affected by something new in a way that qualitatively changes its dynamic nature *relationality*. Relationality is a global excess of belonging-together enabled by but not reducible to the bare fact of having objectively come-together."

Relationality in this sense of transcontextual anomaly or excessive belonging-together occurs whenever differences come into play. In the case of wireless networks, the "complaint" about the Wi-Fi kludge is quite specific: it is directed at the Wi-Fi standards, 802.11b (IEEE 1999) (and presumably would also apply to more recent versions (IEEE 2003)). One way of understanding what might be excessive in a standard is to look at how its limits are defined. Without delving too deeply into technicalities here, IEEE Standard 802.11b or Wi-Fi is defined as part of a larger suite of standards dealing with digital communications in networks that use packets to transmit data, the IEEE 802 family. Other members of the 802 family include WiMax, Bluetooth, and Ethernet. These interlocking standards, usually implemented in computer code, sometimes built directly into semiconductor hardware, form the fabric of the Internet. The standards document for 802.11b published by the IEEE is titled "Wireless LAN Medium Access Control (MAC) and Physical Layer (PHY) Specifications: Higher Speed Physical Layer Extension in the 2.4GHz Band" (IEEE 1999).

Although specifications are intended to reduce ambiguity and eliminate openness to unexpected interactions, sometimes they have the opposite effect. As the title of the IEEE document states, the 802.11b standard describes a way for computers to be networked together using an unregulated portion of the electromagnetic spectrum, 2.4GHz. This is described as a "Physical Layer Extension." Like most contemporary standards and protocols, the 802.11b standard is enmeshed in a web of other standards and protocols. In its very title, Wi-Fi standard refers to and relies on a broader model for communications known as OSI, the Open Systems Interconnection model (on the significance of OSI in the history of the Internet, see Abbate 2000, 167–177). In this model (as well as in the main model for the architecture of the Internet, TCP/IP), the phrase "Physical

Layer" designates the physical and electrical components of the computer network. The physical layer is typically understood as the most stable, inert layer of the Internet. It includes all the wires, cables, optical fiber, microwave links, network sockets, and telephone lines out of which contemporary computer (and increasingly telecommunication) networks are cobbled together. Everything else in the network architecture contrives to hide the physical layer, to push it "down" to the bottom of the so-called protocol stack and to literally put it behind walls, in server rooms and/or inside manufactured hardware such as semiconductor circuitry (Smith 2004). Often it remains visible only in the form of the 10BT plug office PCs are connected into.

From the standpoint of the kludge, it is significant that Wi-Fi, as well as several other wireless members of the 802 standard family, straddles two of the seven layers of communication—the physical layer and medium access control layer—defined by the OSI model. The kludge of IEEE 802.11b, we might say, stems from the differences between the amorphous outline of the physical layer, which no longer resides in cables but occupies airwaves, and the organizational topology of the network implicit in a specific medium access control protocol, the Ethernet, which limits the number of networks nodes (or attached computers) and organizes them in a treelike hierarchy suitable for local area networks (such as those found in office buildings). In 802.11b (and its companion versions 802.11a, 802.11g, 802.11n), the physical layer has spread out of cable into the electromagnetic spectrum. Once it moves out of wires into the electromagnetic spectrum, the physical layer crosses some of the social, political, and cultural boundaries aligned with built space (for instance, the line between public and private) that the medium access control layer takes for granted and seeks to hold in place in the form of a network.

In contrast to some electronic products, wireless devices lie at the boundary between network topologies and physical space. Dourish and Bell (2007, 428) observe that there is something "physical" about wireless networks that renders their relation to space more complicated: "One fascinating aspect of the move from the systems we built on the wired Internet to those that we experience through wireless and mobile networks is that we are creating not a virtual but a thoroughly physical infrastructure, and we need to think about it as one that is interwoven with the existing physical structure of space."

The awkwardness of Wi-Fi mobility, and the reason that it generates so many forms, incarnations, experiments, and hacks, comes from juxtaposition of a medium access control protocol meant for well-defined, centrally administered, and self-contained local area networks such as offices with a proliferating physical layer, propagating signals across once impermeable boundaries and between once divided spaces (home office and kitchen). Importantly, the "kludge" that the software developer refers to is deeply entrenched in the standard. It is not an accident that has befallen wireless devices due to bad technical work. The coalescence of divergent organizations of space and movement at the interface between the physical and medium access control layers produces interesting instabilities.

The kludge both suggests why Wi-Fi will undergo constant transitions and why it will generate a variety of mutations, performances, implementations, and instantiations. The abundance of Wi-Fi-related phenomena can be read as animated by an instability arising from competing logics of space, communication, and movement already visible in the architecture of the Wi-Fi standard. At this moment in the supersaturated medium of communication networks, many different imaginings of mobility and connectivity are in contention. They project different network topologies, different patterns of movement, ownership, regulation, sociality, and embodiment. In differentiating the parameters of these spaces, movements, and controls, communication infrastructures become a locus of social-cultural-material struggle. Wireless networks, as material-social-cultural process that changes communication infrastructures on a variety of scales, precipitate a diversity of movements that shift thresholds between public and private, between individual and collective. At the moment, two principal topologies inhere in IEEE 802.11b. The first presents Wi-Fi as a way of combining access to a medium, the Internet, in many more locations. We could call this the "medium access control" idiom. The second regards wireless LANs as a way of making visible, experimenting with, or engaging with certain physical, economic, and even political obstacles at the edge or in the infrastructure of information networks. The later topological interaction might be called the "physical layer."

The broader point is that wirelessness as a mode of contemporary experience harbors discontinuities between different kinds of relations. James writes that "there is vastly more discontinuity in the sum total of experiences that we commonly suppose" (p. 65). Although experience is really nothing but a series of transitions, and one experience can do nothing but lead to another, the transitions and tendencies that constitute experience do not necessarily fit together. James's striking image of living "upon the front edge of an advancing wave-crest, . . . falling forward" (p. 69) does not guarantee smooth travel. The front edge of the wave crest could just as well be the roughest, frothiest, noisiest place to be, the place where many different things jostle each other. Different tendencies or paths of transition vie with each other in wirelessness.

Individualizing Network Access

A medium access control or MAC address is in principle unique to a given piece of networking hardware. Every networked device, including every Wi-Fi card, has a unique MAC address. There is a dream of global and permanent uniqueness associated with the notion that each and every piece of networking hardware in use today has an individual numerical address assigned to it. Why should that uniqueness be desirable? While the figure of the network as a flat, nonhierarchical, lateral flow has become a quasi-ontological norm in the makeup of business, organizations, groups, and structures of many different kinds, the practical management of networks has a fine-grained governmentality associated with it.[5] An individualism of the network device closely complements the flat topology of the network. The term *Wi-Fi*—coined in 1999 by the Wi-Fi Alliance, an industry association (Wi-Fi Alliance 2003)—is sometimes said to abbreviate "wireless fidelity." The term resonates with *hi-fi*, a term for up-market home audio reproduction technology dating from the 1950s, and tacitly links Wi-Fi to domestic architecture and consumer electronics, to living rooms, televisions, and sofas rather than to cyberspace or office space. This resonance of "high quality but for domestic use" pervades a predominant idiom in which 802.11b equipment and software is visualized, installed, and configured as progress from bad networks to good networks. The transition from bad to good network centers on an individual who would enjoy access to a medium, the Internet, through a device. A powerful vector of the wave front of wirelessness heads toward the promissory horizon of constant, ubiquitous individualized or personalized access to a network medium.

By naming IEEE 802.11-compliant devices Wi-Fi, the industry alliance tackled the representational problem of rendering something visible—a new mode of access to network infrastructure—while stressing its invisibility and its ease of use. However, this implicitly personalized or individualized wireless devices. Network access, or access to the medium, becomes an individual possibility. For instance, around 2002 or 2003, Intel promotions of Centrino™ computing products heavily featured Wi-Fi. Intel is just one salient example among the real plethora of enterprises, schemes, and strategies centering on Wi-Fi as the basis of connected mobile computing. Mobility is understood here as allowing people to more easily use computers in different places by disconnecting computers from walls, wires, and sockets. The Centrino chipsets have been promoted through the slogan "the unwired office starts inside" (Intel Corporation 2003). Integration of wireless capability into the "inside" was represented in magazine

advertisements by an "X-ray" image of the motherboard of laptop computers (Intel Corporation 2003) shown above hotels, golf courses, beaches,
and shopping malls. While "convergence" between communications hardware and computing hardware has been occurring for several decades
(so-called Ethernet NICs—Network Interface Cards—have been standard
computer components for a decade), in producing and promoting the
coalescence of laptops and Wi-Fi, Intel was successful in the ubiquitous
personalization of wireless devices.

Hence, Intel's advertising slogan, "the unwired office starts inside," can
be read as referring to a subjective interiority. The "inside" might also be
that of the "Wi-Fi user," a human subject who has begun to internalize
network connectivity as potentially available anywhere—in public and
private, at work, at home, during leisure, travel, or war. The human
figures—usually men—who populate promotional images associated with
Wi-Fi suggest an "inside" in genesis in two ways. In 2003, Toshiba laptop
computer advertisements showed a man usually alone in remote locations,
although he was occasionally at work in a casually stylish office meeting
(Toshiba Corporation 2003). He stood on a rocky promontory beside a
storm-tossed sea, he sat in a treehouse looking down on the children
playing in a sun-filled backyard, he looked out from a platform high above
a sports stadium, or he lay on the grass in the middle of a park on a fine
day. It was hard to tell who was working and who was not since these men
were not obviously dressed for work. Each time, he looked at a laptop
screen on which some other photographic image had been graphically
superimposed: an office full of people, a library stocked with books, a scene
from an action film. In each case, the superimposed image was somewhat
incongruous with the geographic location. The freedom to connect "in
new places" that Intel's promotions refer to recurs across many different
corporate promotions of Wi-Fi. An affirmation of "freedom"—"enter the
world of freedom computing" (Toshiba 2003), "lose the wires, be free"
(MyZones 2003)—is attached to an absence of wires. Not having to plug a
computer into a socket in the wall to do e-mail, download files, or surf the
Web, means that the screen loses its moorings and begins to float around.
The socket in the wall to which screens are tethered dissolves. In other
words, for the unwired user, the relation between screen and fixed infrastructure changes. Communication is no longer incarcerated, connectivity
becomes quasi-independent of location, and in this liberated space, others
become somewhat invisible.

Yet any attempt to individualize wirelessness as the experience of a
subject accessing a network encounters problems. The commercial promise

of ad hoc wireless attachment to the Internet always had to be accompanied by an effort to make visible places where the networks could be
accessed: *hotspots*. Wireless networks were not, and are still not, everywhere. They tend to be unevenly concentrated in city centers and transport
hubs. The need to make wireless access points visible was recognized in
the corporate strategy of hardware producers such as Intel. Rather than
just including hardware to handle 802.11b communications in its core
chipsets for laptops, Intel "has been working with leading wireless network
service providers, hotels, airports, retail and restaurant chains worldwide
to accelerate deployment and increase awareness of wireless public hotspots" (Intel Corporation 2003). The chip manufacturer wanted to "accelerate deployment" of the technology by negotiating with other businesses
such as hotels, cafés, bars, and airports, and offering a "verification
program": "Intel has developed the Wireless Verification Program, which
includes engineering and testing of Intel Centrino mobile technology with
various access point devices, software combinations, hotspot locations and
wireless service providers to verify they are compatible. . . . The company
expects to verify more than 10,000 by the end of the year" (Intel Corporation 2003). Intel staged "Wireless Days" with free national access in the
United Kingdom and the United States, and also gave awards to cities for
being the "most unwired." Outside the wireless devices themselves, the
hotspot had to become a visible feature in cities, terminals, hotels, and
other facilities. Hotspots were quickly and widely scattered through North
America, Europe, and Southeast Asia (see Wi-Fi Alliance 2003 for a
geographic-location database). As we have seen in previous chapters, they
rapidly multiplied in affluent urban zones such as central London, Manhattan, Seattle, and Singapore, but were also to be found in almost any town
bigger than a village in Europe or North America. Starbucks ("We Serve
More Than Coffee"), McDonald's ("Bites or Bytes, We Do Both"), airports,
hotel lobbies, and bars made themselves into Wi-Fi access points for the
Internet so that drinking coffee, eating burgers, or waiting for a flight
become associated with network access. Flows of food, drink, and passengers merged with flows of data.[6]

Against Individual Access

In principle, at the wireless hotspot, flows of customers, clients, residents,
and travelers register to access the Internet for work and recreation. In the
home, wireless access points connect the entire domestic domain to the
Internet. But networked mobility in streets and buildings also creates

the potential for increased opacity and anonymity. From the standpoint of medium access control or network management, the proliferation of wireless access points has been viewed as a security problem. At industry trade shows such as the "WLAN Event" staged at the Olympia Exhibition Centre in London each year (WLAN 2003), many of the best-attended seminars on the schedule have addressed Wi-Fi *security*. Network administrators and technical information technology directors have regarded Wi-Fi warily because Wi-Fi spreads network topology from the controlled spaces of cables, conduits, and switching rooms. Only after major changes in how users gained access through more secure encryption schemes (from Wired Equivalent Privacy (WEP) to Wireless Protected Access (WPA)) could network administrators in corporations and organizations begin to accept and invest in wireless networks for commercial and institutional settings. From the perspective of the MAC idiom, the world of freedom also means excluding unwanted participants from the networks. Freedom of access comes with freedom from the presence of unwanted others.

The prominence of security as problematic highlights the difficulty in saying who the subject of wirelessness is. Technicians and administrators from corporate IT departments regard Wi-Fi as putting the boundaries of their organization's networks, and in particular, the question of who is inside and outside those boundaries in question. While connections to wires and cable can be visually traced like railway lines, wireless networks spread out diffusely and invisibly, even if they do not go very far. (How far a wireless network can reach depends on the sensitivity of the antennae in use and the local terrain.) The seminars on security, handbooks, and many articles usually figure the "threat" in terms of different possible vulnerabilities and attacks on the integrity of the corporate body. The arrest by the FBI of Wi-Fi hackers in a shopping mall carpark in Detroit (Poulsen 2003), the trial of a hacker who accessed a county court Wi-Fi network in Texas, the largest security breach involving credit card numbers to date (Espiner 2007), or the sentencing of a teenager in Singapore who played online games using a wireless access point in the apartment next door (Chua Hian 2007), all highlight sensitivities about unauthorized access to wireless networks. Unauthorized outside access to the networks is only part of the worry. Danger arises from inside organizations. The software and hardware tools on display at trade shows, and written about extensively in the myriad how-to computer books (Edney and Arbaugh 2004; Barken 2004; Miller 2003), trade publications, and websites, also concern themselves with controlling access *within* the organization. For instance, myriad network analysis tools such as *AiroPeek* allow Wi-Fi network administrators to identify "rogue nodes" attached to their networks by someone *in* the organization as well as blocking attempts to connect to the networks from outside (Wild Packets Solutions 2005)).

Making a Physical Layer from "Virtually Nothing"

In individualizing access to networks through addresses and encryption protocols and pathologizing anonymous, unmanaged access, the medium access control idiom configures wirelessness as individual mobile human users connecting to the Internet. The most important device in network connectivity is the access point that forms the core of hotspots and domestic wireless networks. Quasi-public venues are complemented by a population of wireless devices inhabiting homes and offices. The wireless devices (telephones, laptops, radios, music players, cameras, etc.) found within range of a hotspot/office/home wireless access point offer some release from the postural, gestural stasis of wires, walls, and desks. Yet they are also shadowed by the potential for devices whose operating boundaries are not fully set by medium access control regimes.

Any move away from fixed locations exposes new surfaces where others can begin to appear in modes that are not configured by the medium, the Internet and its dominant access control mechanisms (authentication, encryption, logins, etc.). The physical layer, ostensibly the least social and least tractable aspect of a network, actually can impinge on networks in ways inconsistent with medium access control. Andrew Ross (1991, 98) argued two decades ago in fairly general terms that certain aspects of technology rely on popular participation: "No frame of technological inevitability has not already interacted with popular needs and desires; no introduction of machineries of control has not already been negotiated to some degree in the arena of popular consent." The frame of technological inevitability associated with wirelessness centers on the spread of networks through proliferation of devices: there will be many wireless devices and they will connect everywhere to the Internet (or some version of the Internet). However, this wave of devices advances through "interaction with popular needs and desires" that crisscross between physical spaces and network topologies, between physical layer and medium access control.

It would be difficult to convey the full spectrum of needs and desires around physical space associated with wireless devices. The physical layer has already been discussed in the previous chapter as the air interface. Here the physical layer resurfaces in the many facets of wireless devices that undergo replanning, reshaping, extension, or substitution. These projects span a disparate set of interests, ranging from a geek commitment to exploring the technical limits of connectivity as in the Hurghada project in Egypt (Adly 2003) or the development of a "wireless commons" (WCM, 2003), to UN-sponsored efforts to leapfrog infrastructural hurdles in developing countries (BBC 2003; United Nations 2003). They lack the coordinated global advertising and publicity of corporate promotions. In contrast to the effort to attract individuals to hotspots where controlled individual access to computer networks is available, the common thread in all these projects concerns unearthing communications infrastructures, making infrastructures visible, and transforming them into sites of collective interaction and work. Rather than connecting to the Internet or to the workplace from new places and in new ways, this idiom treats connectivity to network infrastructure in urban and nonurban spaces as holding social potential that goes beyond individuals roaming their own homes, cafés, and hotel lobbies. The physical layer idiom is distinguished from medium access control in several ways: by a nonexclusive relation to others, by some different practices of space and distance, by varying degrees of contestation of commercial ownership of infrastructure, and by an interventionist stance in relation to commodity computer hardware. Potentially at least, this idiom constitutes a metastable, heterogeneous mixture of practices, feelings, and imaginings of communication. A transformation of media-technology habitus, the embodied social knowledge of communication, infrastructure, and urban mobility, could be at stake here.

Let us return a Linksys WRT54G wireless router. Once out of the manufacturer's box, there are no guarantees about what will be done with such a device. Because the Linksys WRT54G architecture is generic and comprises generic components, the software and certain elements of the hardware provided by Linksys can be readily altered or replaced. Two main sites of experimentation can be found in the WRT54G series: the antenna hardware and the firmware.

The device's antenna can be replaced. Antenna modifications probably deserve a chapter in their own right since they embody the infrastructural imaginings deeply associated with Wi-Fi networks. Nearly every website or book on wireless hacking has a section on antennae. Why are antennae so interesting? In the opening paragraphs of the "Do-It-Yourself Antennas" chapter of the book *Wireless Hacks* (Flickenger 2003), Rob Flickenger states:

As you sit at a cafe eating your lunch, you may be completely unaware of the dozens of people simultaneously using the environment around you to communicate with people around the world. I believe that is largely this mysterious, intangible aspect of unseen global communications that draws people to embark on their own antenna projects. The deeply rewarding feeling of making something useful out of virtually nothing is worth much more than saving a few dollars on a network component. (p. 172)

The "making something useful out of virtually nothing" here refers mainly to antennae. In a time when most digital or electronic technology is fabricated in plants in Southeast Asia, the possibility of altering a device using cans, old satellite TV dishes, various pieces of wire, cable aluminum foil, and wire mesh seems for some people "deeply rewarding." It almost seems a privilege to make something. Although many interventions and engagements with wireless networking and digital media more generally focus on making things in software, hardware modifications of antennae in particular provide a stronger sense of agency, a more pronounced sense of making something.

Indeed, antennae become key elements in making wireless devices into wireless networks. *Linksys WRT54 Ultimate Hacking* argues that altering the antenna "can be one of the make-or-break activities that will determine the success of your network. . . . By changing our antennas, we can achieve some very impressive results" (Asadoorian and Pesce 2007, 268). Probably the most iconic modification of wireless routers is the "Pringles Can Waveguide." Although it does not change the power of the signal transmitted, it points it in a narrower beam in a chosen direction, so that widely separated points can be wirelessly linked. By changing the antenna on a wireless access point, the range of the networks can be readily extended to several kilometers, in some special cases up to several hundred kilometers. The upsurge of community wireless networks, municipal or metropolitan wireless networks (discussed in later chapters), and commercial federations of wireless networks such as Meraki (see also chapter 7) and FON (FON 2006) largely depends on substituting different antennae.

Connecting a "home-brew" or separately purchased antenna to a wireless router seems fairly mundane, if not slightly fiddly DIY, work. It hardly seems to invert relations between devices and infrastructure, between experience and the conditions of experience. However, these mundane modifications explore the intersection between the DSP-defined topology of signal envelopes and the Internet protocol–defined topology of the network. Antenna modifications alter signal propagation (longer links, connection through walls, etc.) in the interests of extending a network topology. Changing the antenna changes the range or speed at which information moves. In this sense, it alters the boundaries influencing how the device operates. It affects the kinds of networks of relations that can be imagined between devices. Hence, new forms and distribution of infra-structure can take shape.[7]

"Bricking" the Physical Layer

brick: n. 1. A piece of equipment that has been programmed or configured into a hung, wedged, unusable state. Especially used to describe what happens to devices like routers or PDAs that run from firmware when the firmware image is damaged or its settings are somehow patched to impossible values. This term usually implies irreversibility, but equipment can sometimes be unbricked by performing a hard reset or some other drastic operation. Sometimes verbed: "Yeah, I bricked the router because I forgot about adding in the new access-list." (Raymond, 2003)

Like many consumer electronics gadgets today, a wireless router is actually a computer. As a result, it uses generic chips and electronic components for various purposes. The main onboard CPU, in particular, is nearly always a generic chip since not many wireless router manufacturers can afford to design and fabricate their own CPU. What then of other limits on how the devices operate? Is the physical layer subject to transformation through other changes in the boundaries of the devices apart from the antenna? As we saw earlier, even if we take a single model or submodel of the Linksys WRT54G series, it is hard to even find a single, stable manufactured object. Most of the WRT54G models contain processors made by Broadcom, a well-known designer and supplier of integrated circuits for communi-cations equipment whose corporate motto is "Broadcom in your life: connecting everything®" (Broadcom 2008). This processor can be repro-grammed, and needs in fact to be programmed so that wireless networks can come into existence. In fact, Linksys itself regularly modifies the "firm-ware" that runs on the processors in its wireless routers. Different versions of the same model may have different firmware on them and therefore behave slightly differently.

In a sense, the wireless router manufacturer wants their product to be as solid and reliable as a brick. It should just sit there and help hold some-thing in place, like a brick holds a roof. At the same time, a wireless device cannot afford to behave like a brick because the abstract reality it shares in calls for a degree of dynamism that a brick is not normally allowed to display. So, the firmware on wireless devices always has settings that can be altered or reconfigured to take into account the different situations the wireless device might find itself in. Despite their relatively inert appearance as black, blue or silver boxes, wireless routers have hundreds of settings ranging from highly technical details of transmission power and error correction to security settings, access details for users, and administrative procedures for logging network traffic. The firmware usually supports a miniaturized organized network management infrastructure and puts it at the disposal of the power-wireless user. Typically, these settings can be accessed through a Web interface run from a small Web server on the device itself (see figure 4.3).

In many forms of consumer electronics (set-top boxes, digital music players, game consoles), firmware is hard to alter. Manufacturers regard it as part of their product. Unlike application software on personal computers or mobile phones, changing the firmware, by for instance, replacing it with some "third-party" firmware, voids the warranty. However, like the hardware components themselves, firmware can be generic. For several years, Linksys WRT54G series used a version of Linux, the free-open source operating system kernel, as firmware. Effectively, all Linksys routers were Linux-based computers. Using Linux-based firmware saved Linksys the costs of developing their own firmware, or paying license fees for some other firmware. But generic firmware renders devices alterable. Many people understand how Linux works and many software tools have been developed for tinkering with and adding to it. Subject to software development techniques, the boundaries of the device become much more fluid. Hardly surprisingly, Linksys engineers became aware that its wireless routers were the target of many hacks and modifications going beyond the settings provided in the control interface. In 2005, Linksys released specific versions of the WRT54G such as the WRT54GL that are more open to modification. (The L in this model refers to Linux.) The WRT54GL is specifically intended for wireless hackers. But offering a model specifically intended for hackers only makes sense in the wake of the many modifications that hackers had already performed on earlier WRT54G models. These went well beyond simple changes in the firmware settings that alter the parameters of a device (for instance, increasing the transmission power or changing the clock speed of the CPU can speed up traffic on the network in some cases).

Practices of modifying WRT54G routers are widely documented on the Web (Instructables 2009) and in print (Asadoorian and Pesce 2007). The limit case of modification, the most profound transformations of device boundaries, risk "bricking" the router. The worst-case scenario is that a wireless device—Linksys WRT54G, or the iPhone, PlayStationPortable, NintendoDS, for that matter—can be "bricked" if its firmware is altered injudiciously. Bricking brings us closer to the physical layer because it can collapse all boundaries of the device. Short of this point, substitute firmware such as *OpenWRT*—short for Open Wireless Real Time (OpenWRT

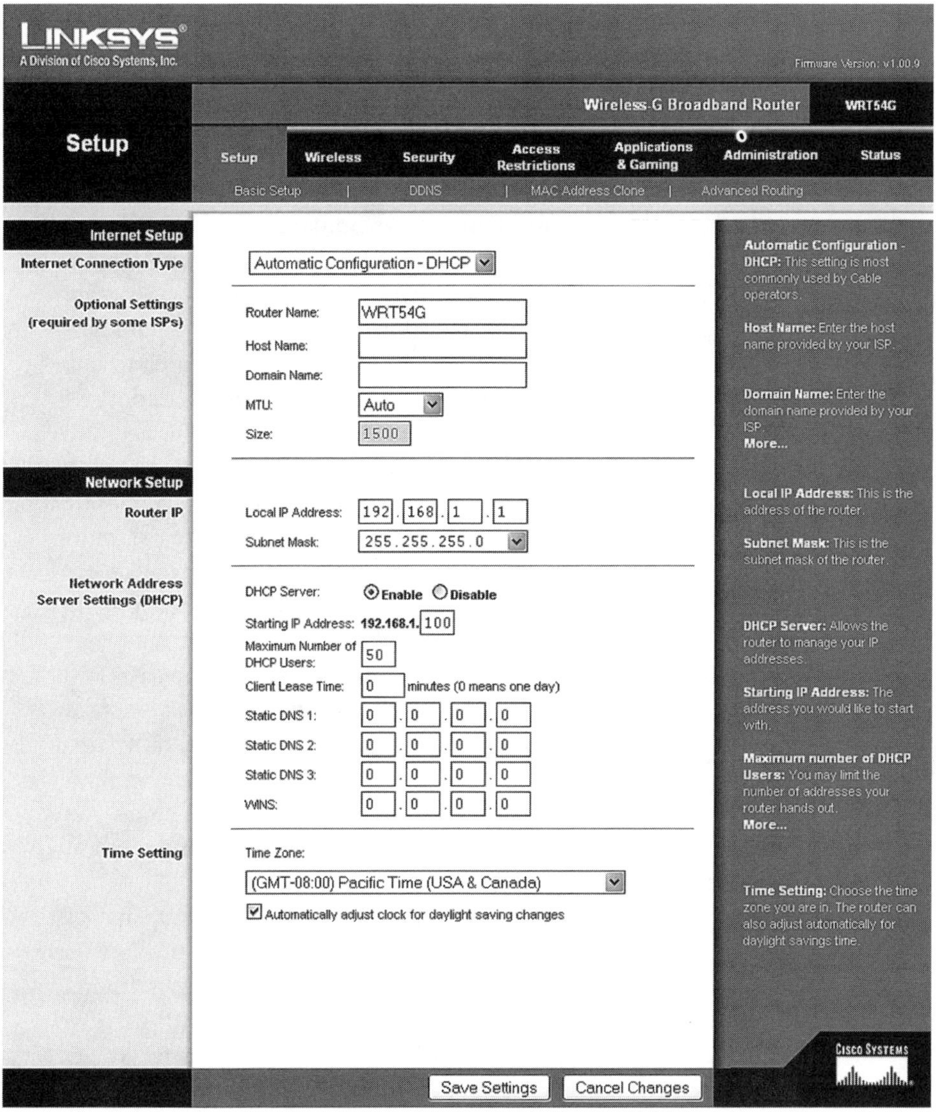

Figure 4.3
Linksys WRT54G Network Setup screen.

2008)—opens space for continuous experimentation with boundaries. As a wireless real-time embedded operating system based on the free open-source software Linux, OpenWRT is a complex piece of software in its own right, with several major versions of increasing technical complexity. OpenWRT offers a substitute for the firmware of several hundred different wireless devices, but for Wi-Fi routers in particular. The possibility of substitution relies on the fact that hardware manufacturers, as mentioned previously, resort to generic semiconductor components such as processors, memory chips, and wireless transmitters. Because the components are generic, the firmware can be substituted. However, this is always a fraught and somewhat experimental process, perhaps much more so than antenna modification, which can be done according to recipes. The possibility of bricking, however, is not the end of the world for OpenWRT. In fact, one of the reasons to install OpenWRT is precisely to "debrick" a device that has been subject to destructive or flawed firmware modifications. But aside from the extremely frustrating situation where some device has been bricked, why would someone bother to substitute different firmware for that supplied by the manufacturer? What explains the desire to reconfigure the boundaries of the device in ways that exceed the configured user already imagined by designers of the control interfaces and antenna?

The OpenWRT project description at openwrt.org claims that "instead of trying to create a single, static firmware, OpenWrt provides a fully writable filesystem with package management. This frees you from the application selection and configuration provided by the vendor and allows you to customize the device through the use of packages to suit any application" (OpenWRT 2008).

As so often happens in wireless technology projects, the OpenWRT project claims a certain kind of freedom, a freedom from the "selection and configuration provided by the vendor." This freedom comes from a substitution. OpenWRT replaces a "single, static firmware" with a "fully writeable filesystem." This means first of all that the device is no longer something that can only be changed at the edges. Once it becomes a "fully writeable filesystem," various points in the device open to intervention. Many of the modifications or hacks described in books such as *Linksys Wrt54g: Ultimate Hacking* (Asadoorian and Pesce 2007), *Wireless Hacks 100 Industrial-Strength Tips & Tools* (Flickenger 2003), or *Wi-Foo* (Vladimirov, Gavrilenko, and Mikhailovsky 2004) rely on OpenWRT to attach new devices (storage, others kinds of network connections), to change the way that the wireless router connects to the Internet, to allow routers to talk to each other, or to change the range of the wireless router by raising or lowering the transmit power of the wireless radio. More exploratory possibilities come from installing software that completely changes the functionality of the device, effectively making it into a different device. For instance, hacks described in Asadoorian and Pesce 2007 can reconfigure Linksys WRT54G to no longer act solely as a wireless router, but also as a DNS server (which resolves hostnames to IP addresses) or as a wireless switch that connects different network segments together. In these kinds of configuration work, the physical layer begins to extensively modify medium access control. The tensions between these different kinds of space can be seen as generative of new relational topologies.

Living between Physical Layer and Medium Access

Work done on the physical layer affects what counts as infrastructure. It experiments with what counts as a good or bad connection, and a good or bad network. It confers the power to speak in the name of a good connection or a good network on different devices or altered devices and hence on different people. From 2002 to 2007, organizations, individuals, and groups interacted with the physical layer in order to enforce or oppose the control and management of network access. They formulated ambitious plans for extending national or international commercial and noncommercial, private and public, local, regional, and occasionally transcontinental infrastructures based on Wi-Fi. Many had explicitly local scope (for example, the wireless community networks discussed in the next chapter, although many of these projects still had forms of global awareness attached). The *Pico Peering Agreement* (PPA 2003) and the *Wireless Commons Manifesto* (WCM 2003) were examples of infrastructure-oriented attempts to move from bad to good networks. These documents represent attempts to engineer connection of local networks into extensive ad hoc informal meshes of wireless nodes across local and national boundaries. The attempts range from manifestos (e.g., "We have formed the Wireless Commons because a global wireless network is within our grasp. We will work to define and achieve a wireless commons built using open spectrum, and able to connect people everywhere"; *Wireless Commons Manifesto*) to quasi-legal agreements that seek to formalize connections between networks (*Pico Peering Agreement*). What would motivate anyone to try to replace international communication infrastructures with infrastructure built and run by relatively ad hoc collectives? Their stance is not simply oppositional. Reporting on a 2003 conference held in Copenhagen to develop and promulgate the *Pico Peering Agreement*, one participant suggests that the consolidation of commercial operations in the 2.4GHz spectrum in the form of "hotspots" in hotels, airports and coffee chains, is not as threatening as it first seemed. These commercial networks continue to focus on wireless network access. The Free Network, as defined by documents such as the PPA (Pico Peering Agreement), has an entirely different and unique potential: to be a viable and competitive supplement to the internet, but one where the system of ownership is decentralised enough for it to remain a "common." (Albert 2003, 7)

These initiatives are directly influenced by free software movements, but take up a political stance in relation to infrastructure based on the proposition that access to communication infrastructure should be free.

Rather than concentrating on hotspots where individuals will access the Internet "in new ways," these projects aim to alter the proprietary status of the infrastructure itself by introducing collectively organized detours, bypasses, and supplements to it. Sometimes they modify hardware or produce software. Typical of the hardware and software modifications, LocustWorld MeshBoxes (LocustWorld 2003) allow wireless nodes to be connected in a "mesh" that can cover an extended area just as a cell phone network with its scattered masts does. As we have seen, antennae are objects of wide-ranging modifications in the physical layer idiom. Antenna modifications that extend the range of 802.11b well beyond the technical limits of a few hundred meters go hand in hand with reconfigured devices that no longer simply provide medium access, but aggregate in different connective formations. Commodity hardware, assembled and modified, becomes part of the practical rhetoric of the cultural inversion of infrastructure. Interactions with the physical layer exhibit a more diverse "sociogeographic" range than those of medium access idiom with its investment in hotspots, homes, and offices. They have a wider geographic range in South East Asia (Jhai Foundation 2003), the Pacific Islands (St. Clair 2003), Africa (Adly 2003), Europe, and the United States, and intersect extensively with development projects (see chapter 7). While the medium access control idiom configures individuals to enjoy a freedom to connect as they commute in and out of conurbations in Europe, North America, Japan, Korea, or Taiwan, collective work on physical layer interactions envisages a different mobility, a mutability in network infrastructure itself. The desire to construct infrastructure, to create a supplementary or alternate "physical layer" for the Internet, is an intriguing and significant development in the post-dot-com cultural politics of communications. It is also a key component in making wirelessness possible. If, as Andrew Ross argues, there is no frame of technological inevitability that has not interacted with popular desires and needs (Ross 1991, 98), then any inevitability attaching to wireless technologies requires something like the physical layer interactions. With its geographic dispersion, its efforts to modify or rebuild commodities (hardware and software) and communities, its aggregation of monadic individual connection into associated clusters or "meshes," and its legal-technical efforts to entrench alternative, large-scale digital infrastructure, the physical layer idiom distends the smooth growth of a network topology in the idiom of medium access control.

Experiments in redefining the boundaries or limits of wireless devices "potentiate" devices as physical layer infrastructure that exists outside or in some respects independently of commercial network infrastructure. Such experiments address powerful enterprises and organizations that corral and tether wireless signals to service plans and network management systems. We could say that a radical empiricist vein of practice runs through these experiments. Radical empiricism supplies techniques for paying attention to the ways pure experience comprises an abundance of connections, tendencies, and differences. Modifications of a wireless device (as well as efforts to resist modification) speak in the name of different styles of connection to networks. Notions such as "autonomy" or "openness" attach alternative sequences or sets of connections to devices, developing their potential to construct alternative network infrastructures. Many of these experiments in resetting the boundaries of devices directly challenge commercial or product-oriented interpretations of wirelessness (as a service or product for personal freedom). The open-ended and sometimes almost practically pointless work done in these experiments is inextricably practical and symbolic, it blends meaning and practice. The work, as Nancy suggests, "remains without end," partly because these devices can only provisionally stabilize some aspects of wirelessness. Wireless connections and the idea of network run unevenly across the wave crests of contemporary experience. Sometimes wireless devices work as components in a stable, taken-for-granted background of expanding networks. Intermittently, they appear as elements of an ecology of excessive belonging-together, overflowing the bounds of sociotechnical network infrastructure.

5 Acting Wirelessly: From Antenna to Node Database

Sustaining, persevering, striving, paying with effort as we go, hanging on, and finally achieving our intention—this *is* action, this *is* effectuation in the only shape in which, by a pure experience philosophy, the whereabouts of it anywhere can be discussed. (James 1996a,183–184)

Consume, a Wi-Fi project active in East London during 2002–2004 and still visible on the Web, sought to add "whereabouts" to the act of connecting to the Internet. Consume's equipment at that time included a wireless access point transmitting from a larger antenna on the roof of the former Greenwich Town Hall, a website representing the current state of wireless connections in a geographic area centered on London (Consume 2003), and a series of public events and exhibition booths based around setting up wireless equipment. The project attracted substantial media attention during 2002–2005. Its motto was an injunction to act: "Trip the loop, make your switch, consume the net." One key figure in Consume, James Stevens, was regularly interviewed by newspapers (Priest 2005). Consume's website at www.consume.net shows a map of London with each wireless access node marked. It provides technical information about how to connect to the node and an e-mail address for each node owner. These wireless access points are scattered across London. In some places their coverage overlaps; in others there are wide gaps in between with no coverage (although again, this depends on the sensitivity of the antennae in use). These nodes are marked as having different operational status—some are active, some are still being set up, some have been taken off air for various reasons.

The Consume project introduced localized, more or less temporary connections between people living in some neighborhoods of London, and relatively short-lived networked connectivity at events in specific places in East London. The "Wireless Clinics" that Consume ran between February 2002 and July 2003 made temporary, local alterations in the topology of networked communications. What did people gather around in these events, apart from the mundane fact of connectivity? Of these events, an observer notes: "Other things that *Consume* does that have been really useful is the events. Sometimes they're more social, fun events where people in a big old town hall use Wi-Fi to download a whole lot of music, and dj" (Simon Worthington, interview, 2003).

Attention seems to move away from wireless technology itself toward the technology as a way to bring people into association with each other without the intermediaries of commercial ISPs or network infrastructures. Projects like Consume are not totally disconnected from or opposed to ideas of network as services. At some point, every wireless network connects back to commercial infrastructures. For instance, on the roof of Consume's workshop in the former town hall in Greenwich, one antenna pointed across the river toward the office towers of the Canary Wharf financial services precinct on the opposite side of the Thames. For over a year, Consume had a 1 Mb/sec Wi-Fi link to a corporate data center there. When the link began to fail (perhaps due to the rampant growth of other 802.11b signals in the vicinity), Consume had to fall back on a commercial broadband connection. Moreover, Consume presented itself at commercial events. In 2003, at one corner of the annual industry WLAN event in London, it shared a stand with the "UK Broadband Consumers" group (ABC 2005). Finally, the name of the project, *Consume,* reflected a somewhat complicated relation to commercial network infrastructures: "In calling it *Consume,* the idea is that it consumes the net, that it should be a replacement for the commercial networks, not just locally but internationally" (Simon Worthington, interview, 2003).

Consume plays on two different senses of the word at once. On the one hand, it issues an injunction to consume. People might be able to consume bandwidth almost for free if they have wireless equipment. They can consume bandwidth, for whatever purpose they can think of—downloading episodes of the UK television series *The IT Crowd* perhaps. On the other hand, the Internet as an increasingly commercial entity to which access is controlled by different thresholds involving payment, will be "consumed" or eaten up.

Information about the locations of the wireless nodes of London displayed on the Consume website is not reliable or accurate since many nodes go off air without any updates to the Consume database. Until someone reports a change in the location or availability of a node, the website will not reflect any change. The primary function of the website consists not so much in providing access to wireless networks. Rather, as one person involved in another London wireless project put it, "Yeah,

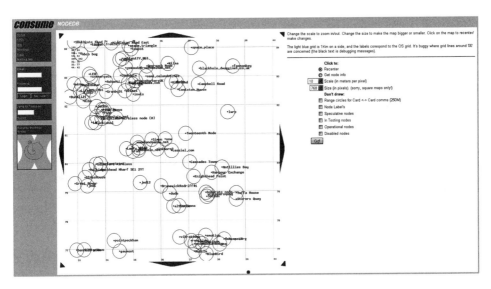

Figure 5.1
Consume the Net, London

. . . it's not intended to be a definitive database. What its meant to be is, in a certain sense, a social tool. You put in your postcode, or somehow you locate yourself on it, and then you seen who's around you based on whatever details they've provided—whether that's a url, what kind of equipment they have. And in a way, you personally make contact with them and see if it's real or not. So it serves some purpose" (Simon Worthington, interview, 2003).

While the Consume database envisaged quite localized and personal contact, wireless action sometimes goes much wider. Projects such as *FON* (FON 2006) and *WeFi* (WeFi.com 2008) display much more global ambitions, and record tens of millions of hotspots (WeFi.com 2008). The forms of acting and the collective practices of belonging, inclusion, and identification associated with wirelessness are divided and in tension with each other. The relation between "connection" and belonging is very unstable in wireless settings, and these instabilities lead to some odd sociopolitico-commercial-technical compromises. These tensions particularly play out around troublesome terms such as *location* or *locative, community,* and more specifically, *openness.*

Wireless action, praxis, and practice are the main concern of this chapter. What does it mean to act wirelessly? It somehow involves connection. The chapter addresses the question of how in the radical empiricist account of pure experience as patchwork of diverse tendencies there could be action, and how in particular wireless action could take place in fidelity to the idea of a wireless network. In particular, in exploring the work done to connect wireless equipment into wireless networks, I am interested in what James calls the "striving" or "hanging on" entailed in wireless connections. The feeling of "paying with effort" is a vital component of the radical empiricist account of action. It stems, first of all, from the radical empiricist emphasis on the reality of felt relations. The feeling of effort also answers to the problem of how to identify an action, given an account of experience that does not fundamentally distinguish thought and things, material and ideas. The key conceptual shift in the account of action proposed by radical empiricism would be away from locating action in subjects toward locating action in the felt reality of relating. James (1996a) says that "sustaining, persevering, striving, paying with effort as we go, hanging on . . . this *is* action . . . in the only shape in which . . . the whereabouts of it anywhere can be discussed" (pp. 183–184), or again, "the word 'activity' has no imaginable content whatever save these experiences of process, obstruction, striving, strain, or release" (p. 167). In both these statements, the endpoint of an action and its point of departure have less importance than sensations of strain, effort, obstruction, and process.

How would such a minimalist account of action constructed around the felt reality of relations play out in analysis of wireless networks? It might, for instance, reconfigure debates around local versus global that habitually accompany network and globalization debates. Tensions between local and global are not new, and they are certainly not confined to the wireless or mobile aspects of information networks.[1] In particular, *connection* to the Internet has an ambivalent status. On the one hand, it makes this place, this here and now somehow less important; on the other hand, it makes it possible to stay in place. In the development of the Internet, the ambivalent status of connection and locality has triggered many debates over identity, inclusion, belonging, participation, and exclusion. However, in their replay in wirelessness, the tensions around the value of connection exacerbate a tendency that has marked the Internet for several decades, a tendency to *expand*. As Tiziana Terranova (2004, 3) remarks, "The design of the Internet (and its technical protocols) prefigured the constitution of a neo-imperial electronic space, whose main feature is an openness which is also a constitutive tendency to *expansion*." The tensions between expansive openness and enclosed control run broadly across many forms of Internet media, and in fact, generate both the heterogeneous forms of activity associated with Web and Internet media as well as their propensity to freeze up in static reductive, habitual forms. As Terranova (2004, 72) argues in her account of a "network culture," "A network culture can never be a unitary formation, describing a homogeneity of practices across a global communication matrix. On the contrary, if such a thing exists, it can only describe the dynamics informing the cultural and political process of recomposition and decomposition of a highly differentiated, multi-scaled and yet common global network culture."

For Terranova, there is no unified network culture as such, only the dynamics informing processes of alteration and mutation. Although there seems to be circularity in this statement ("a network culture . . . can only describe . . . a . . . global network culture"), the central point is certainly important. Network culture only exists as "dynamics informing the cultural and political process of recomposition and decomposition."

The network culture dynamics of recomposition and decomposition can be reframed in terms of radical empiricism by starting again from conjunctive relations, and regarding conjunctive relations as conducive to connection and collection. We have already seen (see chapter 3) that radical empiricism treats the world as a collection of conjunctive and disjunctive relations, concatenated in partial unions through pathways (James 1996a, 107). The practices of collecting elements, finding paths, and creating unions between sets of relations are vital to networks. From the standpoint of radical empiricism, the growth of a network is collective rather than connective in character. Collecting undercuts connecting as the primary practice of networks. A network is a collection of relations. A collection, especially as the term is used currently in software and computer science, lacks any substantial unity apart from that accomplished through "concatenation." Collections may take the form of lists, graphs, trees, sets, bags, or maps. Different kinds of collections cater to different kinds of concatenation and connection. A network is nothing but concatenated conjunctive relations, and any reality of a network as flow, as global, as distributed, as or as collective comes from how these relations are concatenated. The coherence of a network, its ability to transport or sustain flow, depends on the quality and density of those intermediary relations. Many of the dynamics of network cultures can be understood as the interplay of disjunctive and conjunctive parts of the collection.

Given a radical empiricist perspective on networks as collections of relations, how do network dynamics become the *topos* of actions? How do effort, hanging-on, persevering, obstruction, and release weave into the experiential praxis of network culture, into the processes of alteration and mutation, recomposition and decomposition? How are the different forms of being-differentiated, being-multiscaled, being-open, or being-expansive articulated and negotiated in wireless networks? There is no single answer to these questions. This chapter treats antenna modification projects and wireless node databases (such as the Consume project) as technogeographic actions that try to concatenate such paths. Node databases and antenna modification can be seen as experimental efforts to set up loops between networks as collections and experience as path of conjunctive translation.

Where Do the Antennae Point?

Abuse of towers comes as no surprise. (Fuller 2008)

As we have seen in the previous chapters, the very possibility of wireless and hence locative media relies on antennae and radio signals in some shape or form. Signals picked up by GPS receivers, the flow of calls in cell phone networks, or the compression and error-coding processes of wireless networks all depend on antennae being located within range of each other. Range or location cannot be reduced to grid references and map coordinates, especially in cities where signals meet many obstacles. This locative aspect of wireless media—its reliance on the assemblages of antennae and signal-processing algorithms that keep connections in range—is rarely discussed. Does this quasi-invisible aspect of locative media matter? In acting wirelessly, locating, it seems to me, is the principle vector of effort, hanging-on, and obstruction. The act of locating signals, access points, networks, and connection takes multiple forms.

When in the late 1990s, wireless networking became widely available, it was in the form of consumer electronics. Apple Computer's AirPort was intended to connect computers in homes and small offices. A few years later, it became apparent that wireless networks overflow residences. In combination with domestic broadband services, wireless network connectivity quickly began to splash over the confines of walls and buildings, and brim over into streets, parks, yards and neighborhoods. Although the equipment had been designed to cover distances of around 100 meters, people began attaching new antennae that increased the range of wireless networks. Connections over kilometers—and, in some especially persevering cases, hundreds of kilometers—proved possible. These modifications started to subtly change the character of networks in house and business settings. Rather than wireless infrastructure serving as the bridge for short-range connections to the Internet at home or at work, it increasingly began to a form a domain of relationality in its own right. The traits of wireless-ness—its propensities to overflow corporeal, legal, perceptual, and institu-tional boundaries between self and other, inside and outside, public and private—are key to this emergence of wireless networks as a terrain of praxis and commercial enterprise.

One symptom of this is the term *Wi-Fi*. Wi-Fi surfaced sometime around 2002 under the auspices of an industry consortium formed to turn a tech-nical standard into a visible brand (Wi-Fi Alliance 2003). Together with the trademark "Wi-Fi," the Wi-Fi alliance commissioned another trade-mark, the Wi-Fi symbol, that shows an antenna surrounded on either side by waves. The radio antenna, although often quasi-invisible in wireless equipment, is central to any sense of overflowing connectivity associated with wirelessness. The previous chapter discussed wireless antennae in the context of modifications of commodity hardware. Here, in connecting location and the bundles of network relations, antennae have a different significance. What can an antenna do? In contrast to a cellular phone network, satellite devices, or Bluetooth devices, antenna modification is something that is really only easily done for Wi-Fi equipment. Every antenna modification project can be seen as sensitive to location. Location affects the deployment of all wireless networks in very obvious ways. Although the node databases, as we will see, make extensive use of coor-dinate point systems to describe the location of antennae, antenna modi-fication is much more interested in the *elevation* of a location than in its geographic coordinates. Elevation is what allows Wi-Fi signals to stretch greater distances. One recurring trait of antenna builders is their pursuit of rooftops, upper stories of buildings, high towers, and long poles. But at a more down-to-earth level, anyone who uses a wireless device acquires some kind of more or less precise antenna sense as they move through neighborhoods, above and below ground. No doubt, antenna properties are meant to be taken for granted in consumer wireless equipment, since antennae are either invisible (for instance, incorporated in the back of the laptop screen) or can only be twisted or angled in different directions. However, any attempt to render the antenna invisible or forgotten cannot fully succeed. People hunt for access in their vicinity. Sometimes the foot-print of signal propagation draws everyone to one place—the lobby of the hotel, the stairs of the conference center, or more typically, the tables of the café.

The instructions found in the how-to's for modifying or building wire-less antennae are all concerned with sharpening practical senses of loca-tion, and therefore heightening the feeling of conjunction that comes from being better located. The technical knowledges and practices of wireless antenna building focus on the work of bringing locations into conjunctive relation. For instance, the well-known book *Wireless Networks in the Developing World* (Flickenger et al. 2006) describes at length the different properties of antennae and cables. One chapter defines terms such as *directivity, gain, radiative pattern,* and so forth (Flickenger et al. 2006, 88–96). It would be possible to analyze the technical instructions on how to build antennae, showing how they are deeply interwoven with locations—in particular, the problem of elevation, orientation, and obstacles associated with location. The main point here is that every antenna has to take the conjunctive relations that define location into account. (Not only antennae but cables, and as I argue elsewhere in this book, the digital signal-processing algorithms used in wireless networks have to take conjunctive relations into account.) While all wirelessness has this aspect, antenna modification or making projects render it explicit. Even at the level of the antennae, a kind of making of place has to occur if conjunctive relations are to hold, if the world is to have sense.

Take for instance the simplest antenna described in the *Wireless Networks in the Developing World:*

The ¼ wavelength ground plane antenna is very simple in its construction and is useful for communications when size, cost and ease of construction are important. This antenna is designed to transmit a **vertically** polarized signal. It consists of a ¼ wave element as half-dipole and three or four ¼ wavelength ground elements bent 30 to 45 degrees down. This set of elements, called radials, is known as a **ground plane**. This is a simple and effective antenna that can capture a signal

Figure 5.2
Ground-plane antenna (http://wndw.net/)

equally from all directions. To increase the gain, the signal can be flattened out to take away focus from directly above and below, and providing more focus on the **horizon**. The vertical beamwidth represents the degree of flatness in the focus. This is useful in a Point-to-Multipoint **situation**, if all the other antennas are also at the same **height**. The gain of this antenna is in the order of 2–4 dBi. (Flickenger et al. 2006, 95)

It is not necessary to have a detailed understanding of antenna properties to see some practices of location here. First, the very name of the antenna "ground plane" has a corporeal orientation to it. It is not disembedded or space-transcending. It refers to the ground or earth. Second, the design of the antenna has a relation to verticality ("transmit a vertically polarized signal"). Third, the antenna is equally open to "all directions." Although it is open to all directions, it can be altered to provide "more focus on the horizon." Finally, this property changes the relation of the antenna to other antennae, especially if they are "at the same height." Height, horizon, verticality, ground—these directional vectors express basic relations between bodies and things. In terms of location, these vectors define the paths and trajectories of wirelessness. Verticality has a particular importance in relation to elevation and to transmission paths. Although the ground-plane antenna is a simple case, some of the modifications are quite inventive. Already we have seen that antenna design and construction are closely related to geography. An antenna is a geographic-topographic construct. The very calculations that determine how high, how big, and what shape and size it should be, all relate to geography and signal propagation. And this is not a geography seen from on high, but a geography lived as landscape or place, since it is sensitive to obstacles, barriers, heights, depths, nearness, and distance.

The antenna-building projects make visible a significant way of making location today: to act wirelessly is to localize technogeographically. In James's understanding, "Each experience is an action and each action has a center which designates it as a singular reality" (Debaise 2007, 13).[2] The philosopher Gilbert Simondon's notion of concretization provides a way of fleshing out this center of action in the many antenna modification projects associated with Wi-Fi. In Simondon's philosophy, concretization describes the way a technical object or system structures relations between different realities or even different worlds. Although concretization is a complex process of change associated with technical objects, the key point for present purposes is what happens around the technical object. For Simondon, the process of concretization always affects its environment or location. It gives rise to a new kind of location or an "associated milieu"

(Simondon 1989, 55; Mackenzie 2006). Wireless action today, we could say, is making new associated milieus that are technogeographic in the way they relate topography or landscape to network as collection of relations.

In what sense does an antenna modification gives rise to a techno-geographic milieu? And in what sense does this technogeographic milieu allow the object to function technically? For instance, the Wi-Fi modification projects all take place in different locations—urban areas, suburbs, villages, or remote rural sites. We might think that these locations condition or determine the way the technology is made or used. Antennae allow wireless networks to be adapted to new locations. Simondon says just the opposite. Rather than location conditioning or determining the character or nature of action, the act of locating gives rise to what Simondon (1989, 55) calls a "technogeographic milieu": "One could say that a concretising invention realises a techno-geographical milieu, which is a condition of the possibility of functioning of the technical object."[3]

Only in a technogeographic milieu can a technical object function (as node, as access point, etc.). Hence wireless action invents a wireless milieu. Simondon understands a "concretising invention" as hollowing out a relation inside a milieu that it gives rise to.[4] A technogeographic milieu only exists virtually without a concretizing invention that activates practical, semiotic, material dimensions. Localization would be one way of doing this. The process of installing an antenna and logging the position and characteristics of the wireless access point might be understood as "localization."[5]

What milieu only exists virtually in relation to wirelessness? What localization occurs through acting wirelessly? This question cannot be answered simply. First of all, the antenna modification or building projects occur against the backdrop of a highly commodified system of network hardware production connected to globalized telecommunication and Internet service provision, much of which is subordinated to the ideal figure of the network. Second, we usually do not expect to have to know anything about the properties of antennae in Internet media. They are generally taken as given and we are usually discouraged from knowing about them by the complexities of configuration (and this is increasingly the case with the multiple antennae used in more recent wireless devices such as 801.11n). Third, each antenna installation or modification, to a greater or lesser extent, concretizes wireless communication in a specific location. In this sense, although wireless praxis has a constitutively technogeographic aspect to it, its capacity to connect a location is always dependent on other locations.

In the context of network cultures, the technogeographic milieu generated in wirelessness is not just geographical or technical. Or rather, the technogeographic milieu of wirelessness is animated by an incompatibility between inhabited topography and network topology. The antenna modification projects orient themselves toward the collected connections of network. They seek to institute a relation, the wireless link, that forms part of a network. Yet this link is geographically specific. As we have seen, it must take into account orientation, elevation, and what lies between here and there. If there were no node databases, we could say that this link is nothing apart from a slightly enhanced form of concatenation of the world. However, when we view the antenna modifications together with the node databases, we might have a better sense of what the technogeographic milieu of wirelessness looks like, and how acting wirelessly means instituting relations in the center of this milieu.

Locating Practices

If acting wirelessly always passes through acts that make location, or "whereabouts" as James puts it, then one contemporary form of this work is associated with "locative media." Calls for locative media started proliferating sometime around 2002. In broad terms, locative media sought to render networked life and being-connected to networks more habitable, plural, and messy in comparison to the somewhat austere, cool, and anonymous renditions of networks as cyberspace. Much locative media is wireless. In an article that responds to those calls, "Labours of Location: Acting in the Pervasive Media Space," Minna Tarkka (2005, 4) asks "what kinds of potentialities, for thinking and acting, are performed into being in locative media?" Locative media, media that relies on geographic data such as GPS or other ways of marking place to describes its own location, remarkably quickly became a vector of art practice.[6] Locative media also became a highly invested area of new-product research and development in the service of intensified targeted advertising and the provision of personalized content. For instance, many of the free municipal wireless projects adopted business models predicated on the promise of locative media: by knowing where people are in the city, they will offer people free access to the Internet or just the Web in exchange for viewing some location-specific advertisements. These business models have proven extraordinarily prone to collapse and debacle.

Many practices associated with Wi-Fi concern location. *Location*, like *community* or indeed *action*, is a contested and ambivalent term. Hence, the act of "locating" or the status of being "locative" can become a topic of practice. These practices include building wireless networks in particular places, finding out what wireless nodes are present in a particular place, constructing databases of wireless networks, registering Wi-Fi nodes on Web-based wireless network or "node" databases, annotating access points on maps, and modifying wireless hardware (as discussed in the previous chapters). These practices are not necessarily artistically or creatively very interesting. In some of the cases I discuss, these practices are very mundane or explicitly commercial. However, I see these mundane practices involving materials, things, and images as potentially interesting material precisely because they are deeply imbued with effort, struggle and obstruction that link logics of commercial exchange and technosocial praxis.

Concepts of (social) practice implicitly underlie much of the work on locative media. For instance, Anne Galloway and Mathew Ward (2006, 4) see locative media, including wireless network, as able to "to tackle social and political contexts of production by focusing on social networking, access, and participatory media content including story-telling and spatial annotation." In understanding these capacities, they argue for the need to focus on actual practice: "Instead of approaching the physical, the social and the digital as oppositional or complementary qualities, we are interested in how each emerges through the actual practices of locative media" (p. 5). The practices envisaged here concern access, making social networks, storytelling, and effectively sign-posting places for others to move through differently. The concept of practice has been positively valorized in cultural and media studies, sociology, and anthropology in recent decades. A key motif of that work has been the notion that there is no full agent of practices. Agents or subjects are not masters of practices. In various ways they are subjectified in practices. While it would be hard to object to an emphasis on practice without appearing idealist and/or essentialist, ideas of practice sometimes gloss over distinctions between different kinds of action. Older distinctions between poesis, theoria, and praxis, heavily debated in political theory since Aristotle, flow under the notion of practice. Praxis is the most highly valued mode of action in traditional and modern political thought because it is other-regarding. Arguably, an idea of praxis still motivates most contemporary analyses of practice. The philosopher Cornelius Castoriadis (1987, 75), drawing on Marxist thought, characterizes it as doing something in relation to others: "In praxis, there is something *to be done,* but what is to be done is something specific: it is precisely the development of the autonomy of the other or others. . . . One could say that for praxis the autonomy of the other or of others is at once the end and the means."

The development of autonomy of others functions as the axis of praxis. In contrast to the liberal notion of individual autonomy, praxis heightens the autonomy of others. (Poesis and theoria, on the other hand, do not necessarily have the autonomy of others in view at all.) In contemporary network cultures, the question of what constitutes praxis becomes both more difficult and more pressing. On one hand, almost every networked action concerns others somehow. On the other hand, not all networked actions concern the "autonomy of others."

Wireless praxis participates in network relations. These relations are not always compatible or consistent. Wireless praxis, therefore, is likely to be complicated, compromised, or messy. For instance, the ironic title of the Consume project hints that there are conflicting ideas of being a citizen and being a consumer at work around wireless networks. These incompatibilities and inconsistencies in wireless networks are not a limitation or something negative. Rather, I would argue, they signal differences in play concerning who or what is other, and how autonomy materializes amid the network dynamics of composition and decomposition of conjunctive and disjunctive relations. Only amid these differences does there exists any possibility of acting in relation to others. The very possibility of acting wirelessly stems from inconsistency and incompleteness in relations of belonging and inclusion associated with networks. The inconsistency in modes of belonging and inclusion associated with wireless networks becomes more palpable if we compare different Wi-Fi node databases—such as Consume (figure 5.1), NodeDB (figure 5.2), FON (figure 5.3), and JiWire (figure 5.4)—that range between anticommercial and purely commercial in character. FON is a "social Wi-Fi" enterprise that seeks to syndicate hundreds of thousands of Wi-Fi® access points together (FON 2006). It relies on a wireless access node database that covers many different parts of the planet. Both the Consume and FON projects recruit people to register the location and access details of their own wireless access point in a database. Not all node databases allow people to register new wireless nodes. Ji-Wire and many other similar node databases do not allow people to register nodes. But they all allow people to find out what wireless access points are available in a place by searching a database, or looking at a map overlay generated by a database. In comparing these databases, we can ask: How do different node databases condition action, practice, and praxis?

Collections of Nodes and Acts of Belonging

The Consume project afforded a slender possibility of developing the autonomy of others. Each wireless node added to the database slightly developed the range of locations accessible to others. The Consume node database dates from a pre–Google Maps time, and shows the geographic location of wireless access points in central London. People register their own wireless access point and address in the Consume database, and it shows up on this simple map. The project was set up by a very small but quite prominent group of artists-activists in East London and Greenwich, including James Stevens.

The image of Consume dates from 2003–2004. At that time, as the screenshot (figure 5.1) shows in the top-left corner, Consume was making use of a node database and mapping system called *NodeDB*. The NodeDB software comes from Sydney, Australia, and was developed by a community wireless group called *Sydney Wireless*. The NodeDB software was originally meant to facilitate the work of community wireless groups. We can see what the NodeDB software is doing by looking at one of the nodes on the Consume map. Here is an entry from the London node database. Each entry in the database characterizes the location of one wireless access point by generating a list of adjacent nodes:

Name:Mildmay Grove OS Grid:TQ 33118 85083 Altitiude:0 Lat/LongN51:32:54 W00:04:48 Status:Operational Description:Roof access with clear 360deg views. SMC wireless bridge with FreeBSD gateway. No external antenna for the moment but would like to if I can connect to other nodes via MeshAP. Contact me if interested. Draw a map.
Draw a map (using streetmap.co.uk).
See an aerial photo (using streetmap.co.uk). Owner info

People enter some basic information (see table 5.1) about the wireless access points where they live or work. They annotate a fragment of a potential network. Then, for other people, the node database gathers information about where and how to connect to the Internet. The key elements of this network making—the pieces that allow the form of the network to appear—are map coordinates. Either a GPS receiver or an Ordnance Survey map has been used to find the grid reference for every node in the Consume database. The database and Web backend then generate the table of values describing the location and bearing of other nodes. In 2002–2003, the Consume database helped a fledgling, patchy network infrastructure begin to materialize. The infrastructure takes shape as a mixture of different

Table 5.1
Nodes within 4 km of this node

Node name	Distance (meters)	Bearing
Mildmay Grove	0	0
Earl Of Radnor	84	211
burderLodge	424	134
englefield	673	204
EssexRoad	678	218
Southgate Road	900	194
Brad	985	256
ShackleNet	1001	55
raylab.free2air.net	1038	158
kings	1042	158

forms. Some of them are geographic (the map and the list of adjacent nodes), some are concerned with device attributes (the access-point description), and some mix commercial or financial matters with social relations. These forms do not always mix easily. Acting wirelessly in this context means making the location of the wireless access points known to the others through the medium of the node database and the web pages generated from that database. The very existence of the NodeDB system as a way of registering and searching for nodes suggests that so-called wireless networks at that time (2001–2004) were not actually wireless *networks*. They were wireless access points, connected to other network infrastructures (such as broadband or local area networks). The wireless access points largely remain disjointed or disconnected. They are potentially linked by the parameters of database queries that retrieve the node information, and calculate distances and bearings between nodes. Only the Consume website, with its map and geographic coordinate systems, collects and concatenates the multiple nodes.

Setting up an actual link to a node in the Consume database was bound to be complicated. The Consume database did not really undertake to include relevant topographic data about elevation of the access points it lists (compare *NoCatNet,* whose modest goal is "to bring you *Infinite Bandwidth Everywhere for Free":* its node database in Sonoma, California, offers information on elevations of antennae (NoCat 2007)). Perhaps only someone who was prepared to begin experimenting with an outside antenna would have been able to establish a connection. This does not mean that participants in Consume were not acting wirelessly. Perhaps just the opposite. The absence of an actual wireless network was precisely the reason to try to make the scattered wireless nodes visible on a map, no matter how crude. Moreover, if it was not possible to make a connection to a particular node, there was also the possibility of the "personal connection" via the social events organized in South London by Consume. In setting up the node database, in registering nodes, in trying to connect to nodes, wireless praxis precisely concerned the possibility of the development of the autonomy of others, autonomy in relation to the telephone and cable enterprises that run much of the infrastructure of the Internet (for instance, in the form of broadband services).

The same situation, in which there are no wireless networks, prevails today. There are very few purely wireless networks. Wireless mingles with wires. Pure wirelessness is a tendency produced at the intersection of multiple forces. Thus, if we turn to the many commercial wireless networking node databases, there too, wireless networks take a form akin to what we have already seen in Consume—that is, a node database. For instance, one of the leading commercial wireless node databases is JiWire, a "Wi-Fi Finder and Hotspot Directory" (JiWire Inc. 2008).

JiWire offers people help in finding a wireless access point in a given geographic place. By clicking on a map or searching for a place or postcode, people can find a list of locations in the vicinity with Wi-Fi access points. Here location takes on a new value. The locations of several hundred thousand access points that JiWire records in its database are nearly all places of business. In contrast to the Consume database where nearly all the nodes were in houses, apartments, meeting rooms, and studios, the nodes in JiWire belong to hotels, airports, cafés, and in a small number of cases, cities. While JiWire offers free connection information to users, it sells location-specific advertising space to advertisers linked to specific locations—a Hyatt hotel, a Barnes and Noble store, JFK New York or LHR London airports. So something is exchanged when someone locates a wireless node they want to use. Their use of the JiWire website translates into a location-specific advertising mechanism—in fact, a networked form of advertising: "The JiWire advertising network spans premier worldwide locations. Major international airports, hotel chains, cafes and citywide networks all use our unique Ads for Access™ advertising formats to engage Wi-Fi users no matter where they connect to the Internet" (JiWire Inc. 2008).

Although there is no wireless network as such, only a large set of wireless access points, there is an "advertising network" that allows JiWire to "monetize users throughout each Internet session" (JiWire Inc. 2008).

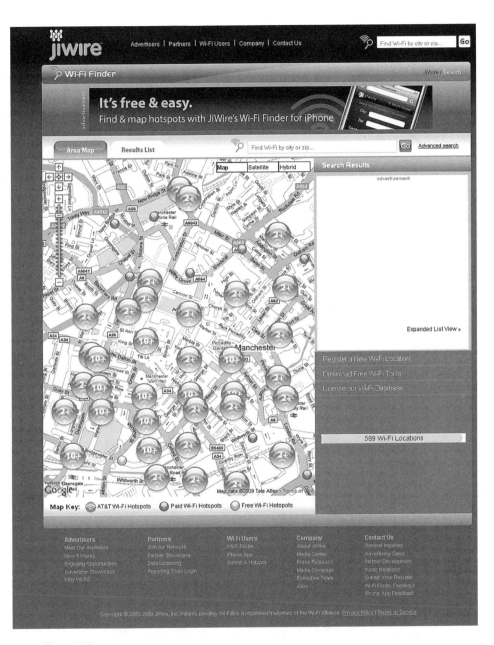

Figure 5.3
JiWire node database

Although people could connect to the Internet through wireless access points offered by JiWire, the network itself was not wireless.

In both the commercial and noncommercial wireless node databases, we confront attempts to collect locations of nodes. These collections grow through contributions, collations, or submissions of information about the location and availability of wireless access points. The node databases, whether in the activist form of Consume or the highly consumer-oriented and designed-up JiWire, attest to the fact there are few wireless networks. This constitutional incompleteness affects what it means to act wirelessly. If the network is a key figure of relational totality today, Internet media are defined as the embodiment of the network to an advanced degree. Any specific medium only shares in the relational promise of the Internet if it too can appear as a network.

Node Database: Wirelessness and the Externality of Sense

The case of Sydney Wireless demonstrates the trajectory of wireless action in rendering wireless networks as networks. Sydney Wireless (Groth 2006)

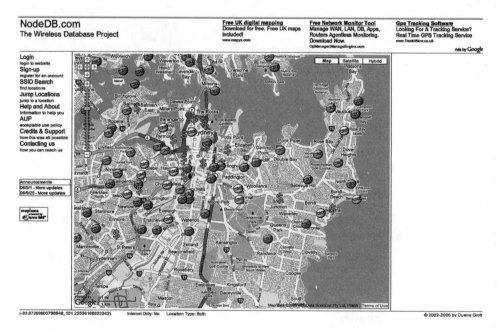

Figure 5.4
Sydney Wireless (NodeDB.com 2007)

is the group of software developers and wireless community activists that wrote the NodeDB.com software used in many wireless networking projects.

As we can see, their node database in 2006 looks very different from the Consume database. It looks much more like Jiwire or FON than Consume. What has happened to NodeDB.com to make it look like Jiwire? Several changes have occurred since 2002 when the NodeDB first appeared online. Community wireless groups often display a somewhat uneasy ambivalence around the relation between citizenship and consumption. While the Consume project's strategy of dealing with this was irony, the change in Sydney Wireless over the last few years has expressed this ambivalence fairly openly. In 2002, Sydney Wireless aimed to "provid[e] a means of by-passing per meg charges and slow uplink connections for people in the community, to play games and experiment with technology that we are otherwise unable to do" (Groth 2006).

Today, providing the means of not having to pay for network connections seems to be only a minor part of the project. As the screenshot shows, many of the nodes entered in the NodeDB.com database are not free wireless access points. Many are commercial hotspots, indicated by dollar signs. This transition in aspirations is described by the NodeDB.com developers in the following way: "What started out as an attempt to map the Sydney Wireless project nodes in Sydney Australia, ballooned quickly into a world wide mapping service for locations the world over. NodeDB.com is essentially a site where people can go to list their location and information about access points or fixed client connections, so others about can easily locate them based on a geographic directory service" (NodeDB.com 2007).

There is nothing surprising about the project. There are others that do something similar (*JiWire*, but also the Wi-Fi Alliance's own Wi-Fi hotspot finder). The *NodeDB.com* sites bears many trademarks of contemporary commercial locative actions. One change is a shift in scale: from Sydney to "the world over." But any shift in scale of action relies on a capacity to map the world. While the *Consume* database could still map London as a scattering of nodes, the *NodeDB.com* site programmers' shift to *Google Maps* as the mapping component of the site is an important change. The freely available *Google Maps* API (Application Programmers Interface) attracts locative media developers, because it is so easy to embed in web pages, allows overlays to be generated easily, and provides free maps for any "website that is free to consumers," as Google puts it (Google Inc. 2008). We could say that the availability of the *Google Maps* API gave a community wireless access project the means of becoming "a world wide mapping service for locations the world over."[7] If the *Sydney Wireless* group started by undertaking a modest form of praxis that developed the autonomy of local others as consumers of Internet bandwidth, their subsequent ambition to become a worldwide project gets entangled not only with the database and mapping services of Google, but with a heightened sense of their own commercial promise.

We Become It

Through a wireless node database, people act on different scales, ranging up to the global or world level. This scale-shifting movement is conjunctive in character. Rather than wireless networks serving as a means of transcending location, the node database places people, as Daniel Miller and Don Slater (2000, 18–19) argue, "in wider flows of cultural, political and economic resources. The boundaries of markets, nations, cultures and technologies become increasingly permeable, and require people to think of themselves as actors on ever more global stages."

Both parts of this formulation, written with reference to the Internet a decade ago, help in understanding ongoing transformations in node databases. We have already seen how the boundaries of technologies become more permeable in wireless networks. Wireless action epitomizes how "increasingly permeable" boundaries "require [that] people think of themselves as actors on ever more global stages." It may be that node databases are interested in, as Julian Bleecker (2006) writes, "elevating . . . geographic locale beyond its instrumentalized status as a 'latitude longitude coordinated point on earth'" precisely because wireless action is on ever more global stages. Viewed from this perspective, the difference between Consume and nodedb.com derives from the frame within which participants are encouraged to think of their actions. To act wirelessly with Consume is different from NodeDB.com. The latter orients itself to a set of external commercial relations that call for a global mapping of wireless access points. The "global stage" in NodeDB.com and in any commercial node database such as JiWire relies on the global mapping services supplied by the Google Maps, itself an Internet service. There is a recursive invocation of networking at work in this concretization, one that gives rise to a technogeographic milieu that links very different scales of action: the few hundred meters of the wireless network access point versus the zoomable, panning view of city and country provided by the Google Maps API.

Staging wireless action globally has a cost. The "we" of the node database can gradually become "it."[8] The organization of action on global scales makes it much more permeable to global systems of marketing and productization, configured to provide services to users. We have already seen this affect Sydney Wireless as it became NodeDB.com. It becomes more obvious if we move to a final species of the wireless node database: wireless social networking enterprises such as FON, WeFi (WeFi.com 2008), and Whisher (Whisher Solutions 2008). FON is "the largest WiFi community in the world. FON is a Community of people making WiFi universal and free. Our vision is WiFi everywhere made possible by the members of the Community, Foneros. We share some of our home Internet connection and get free access to the Community's FON Spots worldwide!" (FON 2006).

FON, based in Spain, after a minor blaze of publicity in early 2006 managed to register tens of thousands of users, especially in Europe and North America. People ("Foneros") agree to open their home wireless access points for free to anyone else who belongs to FON: "Fon is WiFi for everyone" (FON 2006), or at small(ish) cost to anyone who does not belong. Their home wireless access points, as well as a personal profile, then appear on the FON maps produced by the FON node database. It is important to note that FON also supplies modified wireless access-point hardware ("la Fonera"), as well as special-purpose antennae to people who sign up to it ("la Fontenna").

What does this overlay of belonging and technical infrastructure do (see figure 5.5)? FON tries to use the node database system of coordination to create a quasi-global ("WiFi everywhere") collection of access points. Such projects attempt to create a technical infrastructure that can support commercial services by overlaying a social network on it. Like many other social software or Web 2.0–styled entities that attempt to generate revenue by enrolling the work or actions of indentured users, FON uses a social networked form of individual identity to attach a technical infrastructure to places. It thereby generates services sold to other people. Again the node database is the primary site of coordination for the different technical, social, and commercial layers of action. However, all of this occurs in the name of a form of service provision. The product that FON sells is access, but now in a way that mixes the act of belonging and earning money: "FON will pay you 50% of the net revenue that we get every time a visitor purchases a FON Access Pass through your FON Spot! There's more. To help you promote your FON Spot, visitors can get 15 minutes of free access

Figure 5.5
FON node database

to trial the service before they start purchasing FON Access passes from you! And we even pay you for this!" (FON 2006).

Wireless node database enterprises that explicitly juxtapose technical and social networks are apparatuses that capture social relations in the interests of marketing or selling services. Network cultures and informatic formations are deeply imbued with the logics of exchange. In this case, we could say that the node databases and the ideal of universal free access to Wi-Fi everywhere have been transformed into equipment for marketing products and services.

With Others, in Circulation

If resistance is met, *its* agent complicates the situation. (James 1996a, 165)

James's concept of action as a feeling of persevering, resistance, or hanging-on does not specify much about what the goal of action should be. However, construed as a theory of wireless action, that feeling of hanging-on, it should be apparent, concerns connections and making networks. Alex Galloway and Eugene Thacker (2007, 100) ask: "What would a network form of praxis be like?" Their answer to the problem of how to act in a way that develops the autonomy of others is, partly, "write code" (p. 100). In writing code, scarcely visible relations of force can be experimentally configured in ways that spark divergent responses or actions. Code can accrue "praxical" force in imbrication with other processes that in turn are altered and reshaped by technical specificities.[9] Sites of exchange between technical and sociocultural processes always entail praxis, albeit in often compromised, or mixed modes. All of the node databases convey something of the role of code in this respect. They rely on Web and database programming to register, sort, map, and display wireless nodes and networks. However, as an answer to the question of how to imagine a network form of praxis, writing code is ambivalent. Code is praxical and antipraxical.[10] Inherently, wireless action is unstable as praxis. As James (1996a, 152) writes, "As soon as the relations of a thing are sufficiently various it can be sorted variously." If writing code is not the only way to do network praxis, how does this praxis occur in locative media, or at least, in those practices of location developed in wireless networks and wireless connections? Atomically, all practice comprises actions. Practice could be seen as habitual concatenated actions. However, actions are intrinsically composite. This could be conceptualized in terms of embodiment.[11] There is no totality that defines the important of an action belongs. Rather, acting always calls into play relations, connections, and wide-ranging conjunctions that that cross established boundaries. We value certain actions because they make the scope of potential connections and relations more visible. What we call "accountable" action or praxis is a form of doing that affects us, that situates us, and that puts our sense of self in question in relation to others.

Wireless action suffers from constant contamination and contraction of the scope of action. This chapter has tracked how wireless action shifts across different node databases, drawn from activist, community, commercial, and development projects, as well as some antenna modification projects, in terms of the mutations in coding, configuring, and installing networking software and hardware. From a radical empiricist perspective, antenna modification practices highlight the technogeographic practices of locating signals, equipment, and networks. Similarly, from a radical empiricist perspective, the node databases can all be treated as striving to discern paths or "concatenated unions" among collections of nodes. The wireless node databases elicit, attract, or deflect actions concerned with locations. At the same time, the contrast between the databases brings to light the many inconsistencies associated with wireless action. They enable something like "acting wirelessly" at very different scales.

The practices of setting up, running, and contributing to node databases might seem a long way from "acting wirelessly." The databases help us situate "acting wirelessly" in several different, interlocking scales, ranging from neighborhoods to imagined global networks. What do the differences between them signify? The contrast between them can be understood in various ways and at various scales. They can be differentiated in terms of the kinds of practices associated with them—in particular, the kinds of making, altering, modifying, transforming, or exchanging they involve. Some of these practices are highly technical (software development, antenna construction, etc.), others are social (organizing events, attracting media attention, coordinating groups of people), and others are economic (processing transactions, competing for customers, creating brand awareness). These practices, whether they are understood as technical, social, or economic, all center on a certain experience of being-in-wireless-relation. However, across these different scales, something happens to action. Antenna modification and registration with FON entail very different paths of action. While the role of markets, products, and services in wirelessness is the topic of the following chapter, it is evident that something like a capture of action occurs as we move from Consume or antenna modification projects to JiWire and FON. The capture of wireless action is not parasitic on wirelessness. It is not as if there is an authentic action, initiated in the interior of a given actor and affecting others, that FON or WeFi capitalizes on. The radical empiricist account of experience implies no such interiority.

Wirelessness is deeply interlaced in externalized networks of relations because it exposes disjunctive (what is not connected) and conjunctive (what is somehow connected) relations, and develops intermediate pathways or rapport between them. Here, again, radical empiricism suggests the possibility of developing a somewhat different perspective on action. If conjunctive relations are not added to subjectivity, personhood, or things, it is because they comprise the multiple togetherness of things, experience, and others. How can we think about this togetherness without resorting to underlying forms of identity or agency? Nancy's analysis of being-with offers a related take on the conjunctive conditioning of action amid network capture. In *Being Singular Plural* (Nancy 2000), he argues that capital inexorably captures signification and meaning-making processes, leaving only the irreducible relationality of being-with, the irreducible fact that to exist is "strictly inseparable, indiscernible from the *cum* or the *with*" (p. 61).[12] Where in this extension or spacing of relations "with" (which I see as epitomized in wirelessness) does action occur? The historically ongoing process of stripping back meaning or interiority implies that meaning and action never survive intact, or transcend networks of externalizing relations (Ross 2007, 145). Action and meaning retain no footing inside subjects or in belonging to community (even though both the subject and community have often been invoked in resistance to capital). Any sense of "us" or "we" unfurls in stripped-back external relations that affect us ("*our* web or 'us' as web or network, an *us* that is reticulated and spread out, with its extension for an essence and its spacing for a structure"(Nancy 2000, 8)). Since spacings and reticulations of the conjunctive relation "with" are the only irreducible trait of collective life, we need to pay attention to the making and practice of such relations. (If such relations are taken as simply given, we are constrained to live as "windowless monads" (Ross 2007, 148). In that case, the totality of sense will be supplied by networks of external relations grounded in exchange. Relations of exchange will expand globally and intimately. Wireless networks and locative media will be vital vectors of expansion, and will help mythologize capital. Even community wireless networks that focus on the provision of bandwidth to a few dozen local inhabitants may well be aligned with that totality of sense. No authentic community or site of meaning could in principle or in fact resist that alignment.)

Nancy suggests a different analytical path that does not seek to simply substitute a different origin of sense (in the self, in interiority, in the other, in the "we"). Rather he suggests that we need to attend to how the relations are practiced. Sense of meaning originates in the praxis of coappearing, in making networks of external relations. While the point of departure of Nancy's work lies a long way from James's pragmatism, both Nancy's notion of being-with and James's concatenated paths of conjunctive relations center on the praxis of relations. The node databases collect places, things, and people that remain external to each other. The maps with their overlays suggest many possible movements, trajectories, and forms of belonging. Node databases attempt to make visible on the Web the geographic, technical, and social locations of wireless network access. This attempt can never be entirely successful. Indeed, the node databases also make visible the fact that wireless networks are fragmentary, incomplete, and highly transient sites of network connectivity. They show that wireless networks do not remain networks for long. At most, they are collections of access points whose consistency or relationality relies on local orientations and elevations. One way to frame the contrasts is in terms of the intrinsic heterogeneity of networks. Galloway and Thacker (2007, 34) maintain that "the network contains within it antagonistic clusterings, divergent subtopologies, rogue nodes. (This is what makes them networks; if they were not internally heterogeneous, they would be known as integral wholes)." As we have seen, each of the sample databases contains a variety of different network forms and different ways of coping with the inconsistencies and incompleteness of the network. Commercial, technical, social, and political forms mingle quite freely, and are sometimes coupled to each other deliberately. One way to frame their differences is to situate them in terms of wireless action. Following James's radical empiricism, we could say that any sense of acting wirelessly stems from the fact that experience is a "member of diverse processes." The mixture or inconsistency of experience makes action possible and gives rise to practices. We only act insofar as something is not quite right or does not fit, or there is a problem, an obstruction, or resistance.[13]

In this chapter, I have argued that acting wirelessly means locating connections via technogeographic practices such as antenna modification and registration in node databases. The node databases do this in different ways. In some ways they seek to create a technogeographic milieu in which a network could emerge from connections. The node entries in Consume provide instructions on line of sight, elevation, and bearings that help to connect. The modified hardware does this in a different, yet complementary way. The technology itself is modified with new antennae that allow access points to connect to each other, to create new "multipoint situations. Adapting Gilbert Simondon's account of technogeographic milieus, we should understand the node databases and antenna modifications as give rising to a milieu or environment in which wireless connections concretize. Many of the shifts between the different databases could be seen as concretizations that generate and structure milieus of action. All of this seeks to actuate a network.

A technogeographic milieu comes into being alongside the technical object. Following Miller and Slater, I argued that wirelessness, like other Internet media, forces people to think of themselves and their actions on ever more global stages. The increasing presence of world-spanning maps based on the Google Maps API is one symptom of this. The overlaying of social networks on node databases in enterprises such as Fon, WeFi, and Whisher is another. But on these increasingly global stages, the scope of action becomes unstable and open to question. It is no longer simply about where or how to connect. This point allows us to think of the node databases as presentations of action at different scales. Nancy's account of praxis is interesting because it suggests that the most bare, external collections of relations still generate sense. Wireless networks could be seen as literal instances of the external relations Nancy attributes to the working of capital. His account of capital as externalizing networks of relations prompts us to begin to situate wireless action differently. Wireless actions can be seen in terms of a making and praxis of the conjunctive relation "with." As node databases and antenna modifications externalize conjunctive relations, the question of how to make sense through such external networks is the topic of the next chapter.

6 Sorting Inner and Outer: Wirelessness as Product, Wirelessness as Affectional

pay attention to this chapter...

"Outer" and "inner" are names for two groups into which we sort experiences according to the way in which they act upon their neighbours. (James 1996a, 139)

The conjunctive relations processing in wirelessness includes the enveloping patterns of signal processing in wireless chips, the tendencies to vary and proliferate devices in reordering spaces, and the attempts to act wirelessly by collecting connections together in node databases and extending connections through antenna and other modifications. At the same time, from the centers of envelopment of signal processing to the public node databases, wirelessness is heavily saturated with economic and market processes. There is no part of wirelessness disconnected from markets, market research, marketing, advertising, buying, and selling, or not charged by an affective yet banal buzz sourced in capital. Wirelessness seems unimaginable without products in markets. At almost every level from hardware design to network service management, commercial processes and market possibilities animate wirelessness. Wirelessness is thoroughly entangled with products and promises of economic value. The proliferation of Wi-Fi networks certainly changes how digital data and people move, but it is not obvious how these changes relate to vision statements, press releases, media reports, government legislation, certification schemes, online discussions, and advertising in magazines, on television, and on billboards. The fact that specific wireless devices quickly fade into banal backgrounds does not mean that these promises disappear, only that they reinvest elsewhere.

James's radical empiricism provides a way of reconstituting an account of experience around transitions, confluences, and separations of varying degrees of intimacy. Does it offer ways to also explore how the conjunctive relations enmesh with the dynamics of capital? As I have already suggested, because radical empiricism attends to transitions so closely, it is well adapted to making sense of frequent change and the uncertainties associated with the transients and trending of capital. However, in this chapter I want to focus on a single trait of conjunctive experience, and explore how it can be wielded as a radical empiricist technique. One of the "central points" of James's radical empiricism concerns the equivocal character of experience in relation to "inner" and "outer." Inner and outer, thought and thing, are part of the same surface, the same plane of conjunctive relations. Differences between inner and outer, between what seems to be in us and what seems outside us, depend on how this plane is folded or warped. This folding can be understood by analyzing how particular experiences act on neighboring ones, and by tracking the specific process of assembling neighborhoods of conjunctive relations in channels, articulations, or layers of experience. The process of *sorting* of inner and outer is consequential in particular ways. For instance, it allows James to say that ideas of things are not essentially different from physical realities, from something that is, for instance, (experienced as) hard. As James (1996a, 152) says, "These words [*inner* and *outer, mental* and *physical*] are words of sorting." For him, the differences between an idea and a thing depend on the ways an experience acts on its neighbors, as well as how, when, and by whom such an experience is "sorted." An inner experience is not affected by its neighbors in the same way as an outer experience is. In certain respects, it is less affected, since it might easily blend or merge with its neighbors in ways that an outer experience of something hard might not.

Some experiences remain unsorted. Affectional experiences are difficult to sort. Affectional experience has a certain primacy in James's thought because it seems to be run across inner and outer. From this perspective, even minor affectional experiences in wirelessness such as enthusiasm, pleasure, desire, banality, boredom, or fear are worth attending to.

Because it suspends oppositions between internal and external or inner and outer, radical empiricism is capable of remaining open to "extremely complex reticulations" (p. 140) of experience. The complex reticulations that radical empiricism countenances open up a different way of thinking about the promissory economies of wirelessness, and about its existence as a *product* inhabiting a world of networked markets, goods and services. While much of the preceding discussion has tended to treat wireless networks as projects, ideals, devices, and so on, they most often exist as services and products to be consumed. These states are initiated, developed, and tracked by contemporary consumer economies through many different kinds of media. Such states are usually ascribed to a subject of experience. The "wireless consumer" in early years at least, was a male, middle-class, American or European individual. From a radical empiricist perspective, wireless products and their qualities provide an opportunity to explore how lively affectional states enter into the production and consumption of wireless things.

In developing a way of working with affectional dimensions of wirelessness, this chapter scaffolds radical empiricism in three respects. First, it deploys James's account of the sorting of experience as either inner or outer in relation to media treatment of wireless products. The chapter examines how things and people, physical and mental states, are sorted—or not sorted—in media. Practically this means tracking, via a nonexhaustive listing, how Wi-Fi in the years 2001–2004 was treated in different ways in print, online, and in audiovisual media, as either something inner or outer or both inner and outer. Second, drawing on James's account of affectional states, the chapter emphasizes equivocal, provisional suspensions between inner or outer. It explores how the equivocal suspended state of affect entangles wireless devices in the organization of markets around products. This part of the argument draws on the theory of "economy of qualities" developed by Michel Callon, Cecile Meadel, and Vololona Rabeharisoa (2005). Third, any account of the economic life of contemporary wirelessness, it seems to me, must address media processes of subjectification. Here the chapter makes use of the work of Maurizio Lazzarato and Gabriele Tarde in order to understand how subjectifying forces come from and feed back into the economic life of wirelessness (Lazzarato 2002). While William James's radical empiricism and Gabriel Tarde's (1902) "economic psychology" differ in many ways, James's insistence on the primacy of conjunctive relations in understanding the transitions, arrivals, and terminations of experience can be usefully complemented by Lazzarato's explication of how fluxes of imitation and invention constitute desiring and believing subjectifications.

Practically, the chapter is constructed around three lists, three ways of sorting the very large number of media reports, announcements, and advertisements concerning Wi-Fi appearing in 2001–2004. The three lists— one of places, one of devices, one of subject-figures—sort a mixed flow of feelings, practices, and things entangled with photographs, video, articles, and so on. Listing is already writing based on sorting. The chapter traverses the lists, however briefly, not as a series of items, but as "a matrix of immanent universes" (Fuller 2005, 14). As Matthew Fuller (2005, 14) writes, "As a form of speculative writing, the inventory (the list of items and supplies required for an expedition, an experiment, to open up a laboratory) opens up the space of a system of objects arranging itself in composition with as yet unknown combinatorial materials."

These lists derive from scattered encounters with assorted announcements, advertisements, newspaper articles, and online discussions during
those years. They are highly selective, probably not very representative or
statistically valid, samples of a flux of materials that I paid attention to. I
assume that they embody something typical or generic. If these materials
captured my attention, they probably also captured other people's interest
too. The items on these lists were entangled with the flow of beliefs, enthu-
siasms, hopes, and fears concerning wirelessness at that time. While in
other places, I have analyzed such materials in terms of spectacle (Mackenzie 2006, 2005), here something different is at stake. The relation
between items on a list does more than generate associations. As Fuller
writes, "Each element runs off as a gateway to another medial dimension,
an array of processes; each of these is connected up in ways that block,
tease out, evacuate, or mix and amplify potential in what they connect to"
(p. 110). James, Callon, and Tarde all concern themselves with the forms
of relation that engender differences, divergences, and variations rather
than identity or repetition. The lists presented here are partial inventories
that seek to render visible some of the intersections or conjunctions clustered around wireless networks. The lists also associatively embody the
flickers of interest, attachment, detachment, enthusiasm, boredom, and
skepticism felt in seeing, reading, and hearing about wireless projects,
equipment, services, products, and networks during those years. If James
is right in saying that "inner" and "outer" are names that we give according to the ways experiences act on their neighborhoods, then these lists
too are places where experiences act on each other. Here, the inner and
outer aspects of wirelessness were being sorted through a gradual accumulation of connections, influences, imitations, repetitions, and variations.
These lists sample the neighborhoods of interaction, and the sorting that
experiences undergo. Critical analysis of wireless networks should ask why
there are so many different responses, representations, and projects associated with Wi-Fi, and how they relate to each other. To follow the equivocal
grouping of wirelessness as inner or outer, then, would be to connect many
different locations, spatial scales, and boundaries between public and
private, between institutions and commerce, with differences relating to
global-local, individual, and collective identity, political participation,
media, economy, and technologies.

Translocations between Inner and Outer

We "sort experiences," according to James, "according to the way in which
they act upon their neighbors." This sorting is mostly pragmatic. Just as

Henri Bergson would say that perception reflects the scope of our possible movement and action, the sorting of experiences into inner and outer reflects pragmatic concerns. However, sometimes but not always, this sorting is suspended into a sort of metastable state. This happens in "affectional experiences": "With the affectional experiences . . . , the relatively pure condition lasts. In practice no urgent need has yet arisen for deciding whether to treat them as rigorously mental or as rigorously physical facts. So they remain equivocal; and, as the world goes, their equivocality is one of their great conveniences" (James 1996a, 146).

We might understand the ways certain products appeal to pragmatic purposes and emotional interests along these lines. While a product can still urge itself on us as affectionally, there is no urgent need to sort the images, figures, and sensations associated with it as either "mental" or "physical," as belonging to "us" or "it." Something in this state remains "relatively pure" in the sense that it retains "unverbalized sensation." (As James writes, "'Purity' [of experience] is only a relative term, meaning the proportional amount of unverbalized sensation which it still embodies" (94).)

What affectional experiences occurs in the media of wirelessness? Every time an announcement, image, or forecast appears, a claim is made on someone's attention. Any attempt to figure a thing (a product or a service) through images, words, or performances makes connections to other experiences—of home, work, education, travel, war, illness, sport, death or family, government, and so forth. In other words, to figure something means to locate it in conjunction with something more or less adjacent or remote. As we have seen in the preceding chapters, locations are vital in wirelessness, and location above all concerns the position, movement, and habits of bodies. Wirelessness is perhaps nothing else but a series of complex conjunctions concerning proximity and remoteness to bodies in states of tension, movement, sensation, and excitation. Some of these conjunctions can be seen in play in table 6.1. It lists some of the different locations in which wireless connections were to be found in the years in question.

The locations listed in the table are certainly not remarkable in themselves.[1] Faced with this panoply of contexts, it is not possible to make exhaustive claims or conduct multisited observations about what was actually happening in those places. For the most part, it is hard to say what actually happened in the Wi-Fi-equipped truck stops, department stores, caravan parks, or train stations. It may be that for many of the newly networked places of 2001–2004 (apart from houses, apartments, and schools), actual Wi-Fi usage was lower than expected. Regardless, the sheer

Table 6.1
Wireless locations

Apartments	Nations (St. Clair 2003)
Aircraft (Mills Abreu 2003)	Offices
Airports	Parks and squares (openpark.net 2004)
Ambulances	Public events (festivals and shows)
Bicycles (Gitman 2004)	Pubs (Macdonald 2004)
Boats (cruise liners, yachts, ferries)	Reserves (Coeur d'Alene Tribe 2004)
Cafés	Resorts
Camps	Restaurants
Campuses	Rucksack (Mccarthy 2004)
Trailer parks (Glasner 2003)	Shops
Cars (Best 2004)	Showrooms
Cattle ranches (Biever 2004)	Sport stadiums
Gardens	Streets (Kewney 2004)
Gas stations	Trains
Hotels	Train stations
Houses	Truck stops
Islands	Villages (Jhai Foundation 2003)
Libraries	Vineyards
Municipalities	Wildlife preserves (Cohn 2004)

variety of places being populated by wireless networks during 2001–2004 cannot be captured by diffusion models of innovation. Notions of diffusion often appear in accounts of wireless technology. For instance, in their conclusion to *Mobile Communication and Society: A Global Perspective,* Manuel Castells (2007, 258) and his colleagues treat this as a process of diffusion: "Wireless communication technologies diffuse the networking logic of social organization and social practice everywhere, to all contexts— on the condition of being on the mobile Net."

The notion of diffusion does not capture the irregular topologies of the list. How could locations proliferate if not through diffusion? During the years 2001–2004, the idea of data moving along with people was quickly articulated with a number of social, cultural, political, and economic problems. For instance, well-known IT futurologists and commentators such as Nicholas Negroponte and Howard Rheingold placed great stock in the potential of wireless networking to overcome the disappointments of the Internet in the aftermath of the dot-com market crash of 2000–2001, or even to remedy the deficiencies of mass media (Rheingold 2002; Negroponte 2002). In an interview published in the UK newspaper *The Guardian,*

Negroponte predicted: "The current rise of WiFi will . . . bring nomadic computing and broadband wireless to the foreground. When that happens, more and more TV will be nomadic. Basically, it will include everything other than the large format group events. I already look at more TV on my laptop than I do on a real TV" (Gibson 2002).

The idea that the damaging or limiting aspects of mass media can be remediated in Wi-Fi applies by analogy to work and workplaces. Many of the places listed above (hotels, airports, campuses, offices, cafés, etc.) are associated with work. It would be possible to analyze in detail how particular places are constructed and experienced as improved by access to the Internet for work purposes (exchanging documents through virtual private networks, managing e-mail, using the Web, making telephone calls using VoIP, etc.). The main point is that data moving alongside people coalesced with a multiplicity of other expectations, problems, and hopes concerning communication, movement, and location.

Social commentators and theorists also responded to the proliferation of network connections in different places with new theoretical constructs of communicative praxis. The idea of the "Hertzian landscape" put forward in 2003 by the architecture critic William Mitchell (2003, 55) is one such response: "Every point on the surface of the earth is now part of the Hertzian landscape—the product of innumerable transmissions and of the reflections and obstructions of those transmissions."

The Hertzian landscape erases distinctions between places. Mobile phone networks, television and radio transmissions, and wireless data networks flow together to make a "landscape" that overlays and overflows topographic and geographic differences between points. In his account of the networked self in the contemporary city, Mitchell places great weight on certain kinds of networks exemplified by mobile phones and less obviously by the wireless computer networks popularly known as Wi-Fi. What is at stake in wireless networks for Mitchell are transformations of public space induced by the movements of data between mobile points attached to different networks.[2] While the idea of much data moving quickly between points is now well accepted, the idea that data would move in concert with people is still novel and in the process of being articulated in different ideas of mobility. Castells (2004, 87) writes: "Moving physically while keeping the networking connection to everything we do is a new realm of the human adventure, on which we know little."

Wirelessness was articulated as a solution to a problem: how to keep data moving in a way that is synchronized with the movements of a

person. This led in turn to the development of many different kinds of
habits, expectations, and systematizations of mobility.

Finally, advertising and marketing campaigns addressed the problem of
keeping network connections everywhere. For instance, Intel Corporation,
a major manufacturer of wireless networking equipment, promoted Wi-Fi
in the United Kingdom in 2003 using a hot-air balloon in which passengers
rode using wireless-equipped laptops to surf the Internet (McIntosh 2003).
A year later, a Wi-Fi-enabled surfboard (which "lets surfers surf while
surfing") was shown in action at a beach in Devon, UK (BBC News—
Technology 2004). The hot-air balloon and surfboard seem to have been
chosen as deliberately ironic versions of Internet use. At the same time,
the IT press consistently criticized Wi-Fi as the tail end of a wave of hype
surrounding the growth of the Internet during the 1990s : "It was all quite
a wheeze," wrote an observer, "but one that Intel is betting big money on:
the company is spending $300m (£187m) to persuade us that Wi-Fi—or
wireless fidelity—Internet access is the Next Big Thing, rather than a lot
of, well, hot air" (McIntosh 2003).³ Apart from the balloon and surfboards,
it is easy to find examples of hype *and* counterhype or critical awareness
of hype associated with Wi-Fi. Indeed, the IT industry is well aware of the
level of hype associated with mobile Internet technologies such as Wi-Fi.
Gartner Consultants, a major IT industry consultancy, even regularly measures and reports hype: "WiMAX [a longer range, more recent form of
wireless network] hype has peaked" (Smith 2005).

Whenever a proliferation of promises and problems occurs around technology, complicated reorganizations of economic-cultural situations are in
play. The skeptical awareness of the industry pundits and consultants, the
contrived playfulness of the product promotions, and the enthusiasm for
theoretical constructs seem to me to undervalue hype and its dynamics.
Hype is tightly woven into high-tech promissory economies. It has been
analyzed in relation to biotechnology and genomics, pharmaceuticals and
software. For instance, in the context of an analysis of postgenomic biotechnology in North America and India, the anthropologist Kaushik Sunder
Rajan (2006, 113) writes: "Hype cannot be opposed to reality, as is too
easily done when hype is read cynically. Rather, hype *is* reality, or at least
constitutes the discursive grounds on which reality unfolds" (Sunder Rajan
2006, 113).

If hype and reality cannot be opposed, does hype constitute the discursive grounds for the reality of wirelessness? This risks reasserting a problematic primacy of discourse, a grounding of reality on language-based
constructs and concepts. A radical empiricist account of hype would need

to take it seriously, but not reassert the primacy of language or linguistic constructs. The key analytical problem here is: Can a radical empiricism grant hype the reality it deserves? What is the reality of hype then, if it is not a set of statements, utterances, images, and performances that promise something that may not occur? The many images, assertions, events, and documents concerning where Wi-Fi will be tend to offer a diffuse promise. Unlike the promises of, say a biomedical technology such as a vaccine said to have the potential to save lives, or car advertising, with its multiple claims to luxury and status, a wireless network has slender purchase on a subject's personhood and identity. It seemingly cannot promise much. It would be hard to argue that people produce and reproduce themselves (or some aspect of themselves) in wireless networks.

[handwritten margin note: small promise of wifi, low impact on personhood]

So if a list of locations is neither symptom of diffusion nor simply hype, what is it? What if we treated the vision statements, press releases, advertisements, and even social theorizations such as Mitchell's or Castells's as attempts to construct experience in particular ways, classifying them as "outer," as locations? The key dynamic from the standpoint of that sorting is the appearance of a surface of emergence distributed across boundaries between public-private, urban-rural, developed-developing, and individual-social. It seems that during those years, a highly energized process of *translocation* changed the location of networks. Locations were being rearranged, and swapping properties bundled with wireless networks. Taken together, the cascade of announcements concerning these and many other places during those three years reinforced a notion of wireless networking as able to appear anywhere, as ubiquitous and as universal. This did not mean that there was one universal network everywhere. Rather, it suggested that widely socially and geographically variable locations were no obstacle to making networks. Many of them are places associated with transport and movement (aircraft, ships, trains, terminals, cars), but a great many others are scattered in existing urban and rural environments where there are few people, or people do not move fast. The heterogeneous composition of the list (table 6.1) supports the idea that through Wi-Fi, data could accompany people almost anywhere they go. As we will see, the collection of places listed here is equivocally external. It remains open to translocation into "inner," into forms of selfhood. The proliferation of various locations would be a result of that ongoing equivocation. Any vision of a "new realm of the human adventure" would, from a radical empiricist perspective, look rather different. Radical empiricism would replace a database-driven view of "the surface of the earth" Mitchell (2003, 55) with an account of how the "innumerable transmissions . . . reflections

and obstructions" come into relation in order to understand how Wi-Fi
has activated a thousand different projects, reports, opinions, and enterprises of many different kinds. What purpose would this proliferation of
equivocality serve?

Entanglements as Translocations

The list of locations cumulatively creates a feeling of wirelessness as expansive. However, in order to mobilize such a feeling, its status as inner or
outer must remain undecided. We might address the proliferation of equivocality at the level of practical arrangements that entwine wirelessness with
markets. Table 6.2 gives a second, again quite heterogeneous list of wireless
things, this time not of locations, but of equipment that distributes cognition (Hutchins 1995) of wirelessness.

Again, this selective list, drawn from 2001 to 2004, presents a sorting
problem. How can so many different entities related to Wi-Fi all belong
together? Some of the items on the list are products, some are services,
some are expressions or representations of a personal or collective relation
to technology, some are evidence of debates or struggles around the regulation, design, or use of wireless networks. The list runs across production,
distribution, and consumption. It maps out an amorphous and mobile
tangle of relations that can be sorted in various ways.

We could say that this list expresses an equivocal set of entanglements
and disentanglements. In a very literal sense, as we have seen in earlier
chapters, wirelessness promises disentanglement—from the need to connect
with wires, from the constraints of furniture, buildings, or locales, and ultimately, from the unwanted presence of others. Here I am using the terms
entanglement and *disentanglement* directly from the work of the French sociologist Michel Callon on products and markets (Callon 2002; Callon, Barry,
and Slater 2002). One of the key ideas in Callon's work on products and
markets has been that products inhabit markets that are materially structured by dynamic processes. These processes enmesh technologies, sciences
and procedures, institutions and debates. Notions of citizenship, identity,
and consumption entwine with sociotechnical strategies that support competition. However, these disentanglements require other entanglements. In
what sense does Callon use *disentanglement* and *entanglement*? These terms
describe the dynamics of relations between the heterogeneous entities
involved in contemporary markets, with their various forms of regulation,
reflexivity, and above all, with their constant feedback flows across distinctions between production, distribution, and consumption, between prod-

Table 6.2

Wireless things

Access points and routers ("ruggedized," domestic, on a chip, etc.)	Letters and feedback pages in newspapers
Antenna modification projects	Logos and brands (Wi-Fi®, Centrino®, etc.) (Intel Corporation 2003)
Articles (newspapers, magazines, engineering and social science journals)	Modifications of commodity hardware (antenna, mesh boxes, etc.)
Bicycles (Gitman 2004)	Network detection equipment
Billboard advertising in airports (myzones 2003)	Newsgroups and email lists (Hero 2004)
Blogs	Pcmcia, pci, and usb wireless networking cards
Boomboxes (bassstation 2003)	Photographs
Carts (Zuniga 2003)	Protocol documents (IEEE 1999)
Certification schemes	Robots (entertainment) (Williams 2003)
Conferences and conventions (WLAN 2003)	Semiconductor chips
Databases (Consume 2003)	Short fiction and novels (Doctorow 2004)
Digital cameras	Software packages
Domestic sound equipment	Surfboard (BBC News-Technology 2004)
Electronics and computer retail catalogs	Video footage (Gitman 2004)
Forums online	War-chalking marks
Government policy documents	Websites of many genres (wikis, news sites, individual homepages, etc.)
How-to books (Flickenger 2003)	Wi-Fi hotspot access brochures
Industry and government reports (Wolfowitz 2004)	Wireless equipment product packaging
Legislation (spectrum licensing)	Wireless TV (Gomes 2003) (SeattleWireless 2004)

ucts, production, distribution, services, debates, regulation, and identities such as citizen or consumer. In Callon's work, *entanglement* and *disentanglement* designate a much more heterogeneous structuring of products and markets than can be found in most accounts offered either by critical political economy or economic sociology. Rather than alienation, for instance, Callon prefers to speak of entanglement and disentanglement working together. In an interview discussing his own work, he says: "In order to make disentanglement possible, economic agents heavily invest in the production of entanglements. To disentangle you have first to entangle better" (in Callon, Barry, and Slater 2005, 108).

you must have complexity if you want to simplify

In contrast to the classic notions of the commodity as a stable entity that can be transported and consumed indifferently, entanglement and disentanglement occur when different economic actors develop strategies that bring technologies and practices together in the dynamic relations of competition.

A detailed account of the materiality of markets is needed in order to see how wirelessness entails entanglements and disentanglements. For instance, during 2002–2003, as wireless networking products first became visible in many forms, the problem of who could use Wi-Fi for what suddenly became acutely obvious to many people. Ideas of wireless networks ranged from the notion that Wi-Fi afforded a chance to challenge monopolistic ownership of information infrastructures (PPA 2003; WCM 2003; Werbach 2003) to the post–September 11 fear that wireless networks posed a security risk to the State (Boutin 2002). The security of data moving in wireless networks was a particularly concentrated and contested problem. This first surfaced during 2002 in newspaper reports about war chalking, the practice of marking pavements with chalk to show unsecured or accessible Wi-Fi networks in the vicinity (Hammersley 2002). In the aftermath of September 11, 2001, Wi-Fi networks were classified as a threat to national security by the U.S. Department of Homeland Security: "Attention, Wi-Fi users: The Department of Homeland Security sees wireless networking technology as a terrorist threat. That was the message from experts who participated in working groups under federal cybersecurity czar Richard Clarke and shared what they learned at this week's 802.11 Planet conference. Wi-Fi manufacturers, as well as home and office users, face a clear choice, they said: Secure yourselves or be regulated" (Boutin 2002).

hu!

"Securing themselves" began to take various unstable forms of struggle over best practices and modifications to Wi-Fi. In the U.S. military, an order issued by U.S. Assistant Secretary of Defense Paul Wolfowitz (2004) after the invasion of Iraq directed that Wi-Fi and other consumer-grade wireless networking equipment not be used by the U.S. forces, especially those located in Iraq.[4] The order implied that military personnel were acting as consumers, and using consumer-grade Wi-Fi in the field in unpredictable ways. Wi-Fi needed to be regulated for the sake of military security. Finally, during 2003, the government of mainland China issued directives requiring all manufacturers and importers of Wi-Fi equipment to comply with a different proprietary cryptographic protocol called WAPI Wired Authentication and Privacy Infrastructure (WAPI) (Mannion and Clendenin 2003). Because this protocol was designed by Chinese government telecommunications laboratories, the question of whose interests this protocol served quickly surfaced in industry responses: "'The way that they are trying to implement this makes it clear that, whatever national-security argument there may be for encryption, the real motivator is to promote the interests of certain Chinese companies over other companies,' said Anne Stevenson-Yang, the managing director of the U.S. Information Technology Office in Beijing" (Mannion and Clendenin 2003). Several months later, after talks between Chinese and U.S. trade officials, the directive was revoked (Smith 2004).

Attempts by governments to regulate wireless networks tend to make the close coupling between entanglement and disentanglement in Wi-Fi more visible. For instance, in the context of commercial and state concerns about uncontrolled expansion of wireless access, the ethics, politics, legality, and practicality of sharing Wi-Fi networks were extensively discussed. The issue concerned who could access what network. The possibilities ranged from wholesale challenges to telecommunications monopolies through construction wireless networks to the questions of whether it is was legal to share a Wi-Fi network with neighbors. For instance, in the United States, it seemed that

even putting up an unencrypted, unprotected wireless access point might conceivably get you in trouble. Let's say that it's a nice day out, and you want to sit in Riverside park on the Upper West Side and enjoy the day. So you plug your Linksys 802.11(g) access point into your cable modem, and sit outside. You're busted! You see, when you "broadcast" the cable connection, you are opening it up for anyone to potentially use it. So other people can potentially get Internet access from Comcast without paying for it. (Rasch 2004)

In this and many other online discussions, questions about where the boundaries of a wireless network lie were difficult to answer conclusively. Existing legal and contractual controls of data movements come into conflict with new arrangements of people and equipment. While "security" was the major discursive operator in many of the announcements, policies, regulations, and advertisements for wireless networks, counter-vailing liberal ideas of lightly regulated or unregulated movement of data consistently challenged its primacy. For instance, a home Wi-Fi user, Michael Joel (2004), writes: "Last week, I turned off all the security features of my wireless router. I removed WEP encryption, disabled MAC address filtering and made sure the SSID was being broadcast loud and clear. . . . So why am I doing this? In a word, privacy. By making my Internet con-nection available to any and all who happen upon it, I have no way to be certain what kinds of songs, movies and pictures will be downloaded by other people using my IP address."

By not knowing what is downloaded through their own wireless access points, individuals perhaps make it harder for copyright enforcement measures to take effect. On this point, the difficulty of regulating the movement of data between pieces of wireless equipment was entangled with ongoing struggles over intellectual property rights in the "digital millennium."

Singularization and Attachment: The Equivocal Quality of Being Wireless

The notion that every disentanglement requires new entanglements forms part of Callon, Meadel, and Rabeharisoa's (2005) broader notion of the "economy of qualities." This phrase designates a way of thinking about contemporary products as processes or as a series of transformations without resorting to preconstructed macrostructural explanations such as capitalism. The notion of "quality" is quite useful in making sense of the various, often competing claims made for wireless networks. According to Callon and colleagues, markets are organized around trials to establish and test the qualities of products. The qualities in question may be primary or secondary, intrinsic or extrinsic. No kind of quality has any ontological primacy in organization of the market. However, establishing and adjust-ing a list of qualities for a product is a way of positioning a product in the sphere of products, goods, and services at large. The list of qualities is generated through design, through branding, through reviews, or through consumer responses. In turn, market competition today centers on a strug-gle "for attachment [to] and detachment" from products (Callon, Meadel, and Rabeharisoa 2005, 40). The dynamics of the market for services (and increasingly for the service-encrusted goods common amid contemporary wireless media) perform this listing of qualities ("qualification") through two structuring processes. "Singularization" attempts to distinguish prod-ucts from all those that appear similar or identical. This may be effected through the design of the product as an object, through packaging, through "added extras," through placement on the shelf in the shop, and so on. Given that Wi-Fi products during 2001–2004 all claimed to be implementations of the same technical standard (IEEE 802.11b/g), singularization achieved through Wireless Alliance certification and the use of the trademark Wi-Fi were important dimensions of wirelessness. "Attachment" describes processes that actually result in preferences for one product over another, or perhaps for one way of consuming a product over another. Attachments occur through an "apparatus of distributed cognition": "This attachment to a singularized product cannot be disassociated from the configuration—through supply and demand—of an apparatus of distributed cognition in which information and reference are spread out between many elements. The consumer's preferences are tied into this apparatus" (Callon, Meadel, and Rabeharisoa 2005, 38). Attachments particularly seek to bring "beneficiaries" (Callon, Meadel, and Rabeharisoa 2005, 45) closer to the product in question. If a product results, singularization, detachment, and attachment are the processes producing that result.

Viewing the list of wireless things from the perspective of processes of qualification yields several observations. The qualities of a thing are established by mixing materials and mixing practices. Qualities are generated in media in the conventional sense (print, television, radio, the Web); through events (conferences, festivals, parties); in documents, announcements, or hybrid forms (Wi-Fi-equipped bicycle); or via materials and things. These materials and practices crisscross boundaries between production, distribution, and consumption. Much of the print and online media discussion of Wi-Fi adopts familiar formats of organization and circulation of the meanings of Wi-Fi. An industry report such as the Rethink Associates (2005) report consists of a series of product-release announcements, investment opportunities, and summaries of the sales of different equipment and services. Such a report would be read mainly by people involved in the corporate buying and selling of wireless equipment and services. (Indeed, the report itself has to be paid for through a relatively expensive subscription.) By contrast, a report on the IT pages of a daily newspaper is read by many more people but will often focus on a community project in a particular city, region, or place (Wainwright 2003) or on consumer product releases. Here readers are both consumers (they may buy some equipment or service) and citizens (interested in how social problems can be solved).

Qualification embodied in processes of singularization and attachment/detachment continues well after products are sold. Here is an excerpt from a website explaining how to modify a Palm Treo PDA so that it can access wireless networks:

After realizing that the Treo 650 and Tungsten T5 are practically siblings internally, and the T5 already works with the WiFi SDIO card, he [Shadowmite, a hacker] developed a new set of drivers that enable the 650 to work with it. I'm sure the folks at PalmOne, and the wireless carriers will love him for that. VoiP, here we come! No more wireless internet access charges. . . . It's not perfect . . . and the only way to get things right is with a hard reset, but it completely blew PalmOne's garbage press release out of the water. (PIC 2004)

As a form of externalization of meanings, the web page instructions on how to change a device (Palm Treo) through installing extra software (Wi-Fi drivers) alter the qualities of the product as established by engineers, designers, instruction manuals, advertising, and reviews.

While many of the materials listed above are media (photographs, magazines, newspapers, websites, logos, billboards), some are not media in a strict sense of "the media" or even "new media" (websites, e-mails, blogs, digital photographs, video, etc.). A set of chalk marks on the pavement in London, a bicycle taken to public events in New York (Gitman 2004), or a modified antenna made using a Chinese kitchen frying implement in New Zealand shifts and translate the qualities of wireless services and products. They externalize qualities of wirelessness, and differentiate it from other forms of information network. From the standpoint of the economy of qualities, we should think of many of the gadgets, hacks, installations, modifications, deployments, and repackagings of Wi-Fi as forms of "requalification" of wireless networks associated with movements of people and data. Images, reports, announcements, advertisements, and many other things and events associated with Wi-Fi during 2001–2004 were explicitly contrived to make different qualities associated with this kind of wireless equipment noticeably "accessible to the senses." Indeed, a striking feature of the flow of qualities associated with Wi-Fi wireless networks has been just how many different forms they take and materials they use, and the different visibility this variability affords. Some ways of making visible were quite novel, and this novelty heightened their visibility. Wi-Fi projects were persistently associated with developing countries and remote locations (see chapter 7 on this). Reports on these projects often centered on novel combinations of technologies and practices— putting a wireless network on the back of a motorcycle that travels between villages in Cambodia (Chan 2004), extending the range of a wireless connection using Pringles can antennae in Eypgt (Adly 2003), or utilizing cooking implements (Swan 2004).

A second observation that can be made from the list of things is that the process of qualification is equivocal in the sense that James attributes to all experience. It is sometimes difficult to say whether qualities of wire-lessness generated in the practices and things listed there are inner or outer, intrinsic or extrinsic. For instance, while "war chalking" put sets of symbols on buildings, on city pavements, or on websites to alert a passerby that a wireless network is accessible in the vicinity for anyone to use, the only passerby likely to be interested or even aware of the marks would have been someone to whom accessing a wireless network from the street seemed a significant, cool, or subversive thing to do. Similarly, the wireless hotspot signs that appeared in airport departure lounges and many hotels, cafés, and fast-food outlets during these years offered access to the Internet only to travelers who were prepared to pay substantial amounts on a credit card to use the service provided there. (However, see Doctorow 2004 for a quasi-fictional account of the complications and frustrations of access.) While reports of a new product release in a newspaper or on a stand at an industry conference promise new freedoms to connect, instructions on how to modify a PDA or build an antenna that extends the range of the network construct different versions of freedom. In the first case, the promise concerns the intrinsic properties of the device; in the second, it concerns the qualities of the people involved. The process of constructing the list of qualities through singularization and attachment/detachment brings together mixed materials. On certain points, it remains equivocal as to whether the qualities are inner or outer, intrinsic or extrinsic. This equivocation defers any final result or completion of the product as a static entity. It remains open to translocation.

"Everybody's Connecting™": Variations in Subjectifications

James's account of the sorting of experience as inner or outer poses the question: How do forms of externalization of wirelessness in networks populated by devices, products, and services generate variations in subjec-tification, or in experiences of self? Who are the subjects of wirelessness? The final consequence of this equivocality of wirelessness concerns subjec-tification. As mentioned above, Wi-Fi networks in the years 2001–2004 seemed primarily of interest to certain kinds of people on the move (com-muters, business travelers, etc.), people at home, and institutions such as schools and universities. By contrast with mobile phones, whose social distribution covers many different ages, genders, socioeconomic positions, and ethnicities, Wi-Fi connections were tethered to a more limited range of figures. Wirelessness pivots on the hope that information networks might be put into places that existing communication infrastructures find difficult to reach. The variety of locations listed above (table 6.1) indicates that. The variety of wireless qualifications or entanglements (table 6.2) suggests how different locations undergo qualification through processes of attachment, detachment, and singularization. Putting networks in different places immediately raises the question: Who will be there? Many important aspects of the expansion and opening of network infrastructures associated with Wi-Fi can be analyzed in terms of the resorting of experience as inner and outer, or in terms of "which meanings have reached where and when" (Hannerz 1992, 81). In analyzing the sorting of wirelessness as inner and outer, we need to ask how ideas of data moving in different spaces, and externalized in a variety of media and mediations, were themselves figured as matters of subjective experience.

"Everybody's connecting™" is a trademark of Netgear, a manufacturer of wireless equipment. Wirelessness is inseparable from the distribution of ideas and meanings associated with being connected. As a promise to redistribute access to information, wirelessness configures different embodiments of access and connection. For instance, although many projects work to bring wireless networks to places that currently have scant access to the Internet (Waltner 2003; Wainwright 2003; Jhai Foundation 2003; BBC 2003), ideas of access in remote places are mainly distributed among individuals and groups who already have relatively high-level access to the Internet, and who visit websites reporting on the projects. Can variations in subjectification associated with wireless connection be listed in the same way as locations and forms of qualification were in tables 6.1 and 6.2? While it is difficult to represent overall social distributions of meaning, it is possible to outline Wi-Fi as a set of overlapping, mixed, and sometimes competing distributions of ideas of access to networked communication associated with different subjectifications. This leads to a third and final list (table 6.3), a nonexhaustive list of wireless subjectifications.

Different aspects of wirelessness provisionally attached to these figures as they appeared in mass and online media, in advertising and publicity, and in various public forums during 2001–2004. These figures can be seen as temporary, often equivocal stabilizations of attachments/detachments, singularizations, and entanglements associated with Wi-Fi networks in that period. Certain figures here have been the focus of more detailed empirical study by social scientists and market research. For instance, Grubesic and Murray (2004) studied "802.11b activity for one of the more impoverished and violent neighborhoods in Cincinnati" (p. 15), compared it other neighborhoods, and found that affluent neighborhoods showed "a remarkable amount of 802.11b activity" (p. 22). We could imagine other studies

Table 6.3
Wireless figures

Artist (Gomes 2003)	Industry consultant (Smith 2005)
Business traveler (Mills Abreu 2003)	Journalist (Wainwright 2003) (Standage 2003)
Commentator-analyst (Werbach 2003) (Rheingold 2002)	Lobbyist (United Nations 2003)
Developing-world NGO member (Jhai Foundation 2003)	Neighbor (Robbins 2003) (Leyden 2004)
Engineer (protocol, chip, software, equipment)	Network administrator (Flickenger 2003)
Hacker (Swan 2004) (Poulsen 2003) (PIC 2004)	Pedestrian (Hero 2004)
Home user (Joel 2004)	Regulator (Wolfowitz 2004)
Hotspot customer (Macdonald 2004)	War chalker/war driver

of some of these figures as variations in wireless subjectification. For example, during those years, a system administrator installing and administering a wireless network for an office building might have been visiting trade shows (WLAN 2003), reading industry "gray literature" such as *WiFi for the Enterprise* (Muller 2003), and using increasingly sophisticated network analysis and management software (Wild Packets Solutions 2005) to analyze whether or how unauthorized wireless access points were affecting the organizational networks. Today, as fluxes of desire, belief, and expectation wireless networks have shifted, the figure of the network administrator might not be invested in the same ways. (On the other hand, the proliferation of wireless devices means that many people find themselves doing forms of network administration.)

How should the list of figures be understood from the standpoint of radical empiricism? We have seen on several occasions that "holding fast" to conjunctive relations, and particularly to the conjunctive relation of continuous transitions, is axiomatic to radical empiricism. James (1996a 48) writes that "holding fast to this relation means taking it at its face value, neither less nor more; and to take it at its face value means first of all we take it just as we feel it, and not to confuse ourselves with abstract talk *about* it."

How would one "take it [continuous transition] just as we feel it" in relation to the figures of pedestrians, hotspot users, or network administrators? How could one avoid the confusion of "abstract talk *about*" a relation? Is James calling for a more introspective approach to experience? Perhaps he is, but it is possible to use the technique of "holding fast to the relation just as we feel it" slightly differently. We can treat it as a way of attending to the processes of subjectification that accompany the assorted entanglements, attachments, or suspension of definitive categorization of inner and outer. These figures can be seen as points of subjectification, sites where practices concerning ethical substance and technocorporeal techniques attach and take shape. Nobody durably or fully embodies any of these figures. However, many people pass through such figures. Maurizio Lazzarato (2002, 142) usefully observes that "it is in the metamorphoses and variations of action of subjectification and not in the metamorphoses and variations of value that one must look for the immanent dynamics of capitalism."[5] One way to understand the list in table 6.3, then, is as variations of wireless subjectification. While, like Callon, I would be hesitant about reading the list as a direct expression of any macrostructure called "wireless capitalism," variations and metamorphoses in the action of subjectification do occur wirelessly. They are indissociable from the process of qualification that Callon ascribes to contemporary products because these products depend on attachments and detachments that act as subjectifying forces.[6]

Lazzarato's claim appears in his discussion of notions of subjectivity, action, and work in the nineteenth-century French sociologist Gabriel Tarde's "economic psychology." Like Callon's work on products, the model of subjectification developed by Tarde does not oppose ethics and economy, labor and action, instrumental and communicative action, or even work and leisure. (Much social theory maintains such oppositions in the service of notions of freedom, democracy, or ethics.) Tarde, according to Lazzarato, puts these oppositions in question by affirming the force of feelings, affects, beliefs, desires, and passions in constituting not only the experience of individual subjects, but in the very process of subject formation (or subjectification). Like conjunctive relations in James, the forces of subjectification in Tarde are strongly marked by impersonal, prelinguistic, pre-subject-object, and nonrepresentative attractions and repulsions, affirmations and negations (Lazzarato 2002, 123).[7] Tarde brings to the foreground variations and metamorphoses in the "action of subjectification," and draws our attention to the dependency of the value (economic, aesthetic, cognitive) of wirelessness on these variations and metamorphoses. In Tarde's thought, forces of subjectification work via the "affectability" of feeling. In a formulation highly resonant with James's treatment of "affectional experiences," Lazzarato (2002, 21) writes: "It is through affectability, through pure feeling, the common base of desire and belief that Tarde describes the differences of nature and degree between activities of creation and reproduction."[8]

Attending to fluxes of subjectification associated with wirelessness shows that there is no simple sense in which Wi-Fi carries desires and beliefs about communication, connection, networks, place, or others. In any case, it is unlikely that any single or consistent force of subjectification is either reproduced or created in wirelessness.

In the years between 2001 and 2004, when Wi-Fi appeared as a distinct technology (precursors existed for a number of years in various specialist applications), ideas of who could connect to what circulated unevenly. Crucially, however, these ideas acted as desires and beliefs that supported fluxes of practical imitation and practical invention. For instance, one dense concentration of Wi-Fi practice clustered around a particular wireless figure, the hotspot user. While much wireless networking was occurring outside or around the commercial hotspots (in homes, institutions, etc.), the notion of accessing the Internet from a laptop computer at a Wi-Fi hotspot frequently appeared in product and service-provider press releases, images, and articles aimed mainly at corporate and business travelers in transport hubs, cafés, and city centers. Hotspots might in principle be located anywhere, but typically they were found in places where people go to eat, drink, or wait to go somewhere else. In these public or quasi-public places, the difference between being connected to the Internet and not being connected is not merely a matter of having a wireless device or not. It is marked by an altered sense of time that cuts across work and leisure. The act of connecting to an airport wireless hotspot implicates people in a process of service access that reframes the time of departure. In particular, wirelessness at this time engaged people as they arrived or departed, or as they worked away from offices (Forlano 2008; Hampton and Gupti 2008; Laurier 2004). Although the economic value of hotspots was highly contested (different "business models" competed), the conjunctive feelings between arrival/departure and network connection propagated the figure of the hotspot users in various forms. To have a laptop, PDA, or Wi-Fi-enabled phone (and often a credit card) ready presupposed a series of conjunctive relations in place. These relations predicate an individual who pays for network access in advance or on arrival or departure. The provision of wireless networks in "premium spaces" such as first-class cabins on trains and planes and in business-traveler hotel chains expanded widely at that time. At times, a single individual could be figured in relation to different wireless subject positions simultaneously. When wireless services were laid on top of each other, it was meant to furnish a plenitude of connectivity at all times for the never-not-connected. For instance, T-Mobile, the mobile telephone operator, made an announcement in June 2004 that it was offering a combined mobile phone, 3G, and Wi-Fi access package in the United Kingdom to high-value customers: "The company is expecting to attract new customers from both business and consumer markets, but the data card is firmly targeted at the business market" (Kotadia 2004). (This has since become much more common, as reflected in the multitude of devices that bundle Wi-Fi and 3G mobile phone connectivity.)

Conversely, at the same time, the figure of the wireless citizen appeared in highly frequented or symbolically invested locations. Hotspots that epitomized "publicness" were set up, for instance, in a public parks such as Bryant Park, adjacent to the New York Public Library in Manhattan (Bryant Park 2003), in the National Mall in Washington, D.C. (OpenPark. net 2004), and in Picadilly, London. A government minister might see libraries as ideal locations for open-access Wi-Fi hotspots ("I'm very keen on the idea that every public library should be a Wi-Fi hotspot," said Stephen Timms, UK Minister for eCommerce and eGovernment (Kewney 2003)). (By contrast, the local police might see a library hotspot user as suspicious. For example: "A few minutes ago, a police officer passed the bench where I was sitting outside the [Nantucket] Atheneum, enjoying the mild temperature and the wifi signal, and he said, 'Sir, you can't use the Internet outside the library'" (Akma 2004).) Public hotspots in parks and libraries figured people as citizens rather than consumers. Belief in the possibility of hotspot connection to the Internet in all these places was supported by the feelings of conjunction that mixed together consumption and citizenship, work, and leisure. Hence an individual can at one point be a consumer and at another a citizen.

Finally, wirelessness also produces figures who are either problematic or pathological. A person who left their home wireless network open to their neighbors may have seen themselves as neighborly, whereas a broadband supplier might see them as telecommunications felons (Cohen 2004). Similarly, a Sydney man who put a wireless access point outside his suburban house was apprehended by police after reports from his neighbors. They suspected he was a terrorist and that the wireless access point was a bomb (Leyden 2004).

Like the other two lists, there is nothing exhaustive about the list of wireless figures in table 6.3. The figures presented here are points of intersection or provisional sortings of conjunctive relations. They carry forms of wireless affectability. The figures circulate in a great variety of forms of media, ranging from the instructions on the back of a wireless router box to government telecommunications regulations. They are not full or complete subjects, but fragmentary components of existing identities, subject positions, and forms of lived experience assembled in relation to network connectivity.

Recursive Sorting of Inner and Outer?

Thought and thing, subject and object, are not separate entities or substances. They are the irreducibly temporal *modes of relation of experience to itself.* (Massumi 2002, 108)[9]

The sortings of locations, things, and figures specific to wirelessness both maintain and equivocate differences between inner and outer on many different scales (home, neighborhood, city, country, etc.). We have already seen in the previous chapters some erosion of distinctions between infrastructure and superstructure, between network and node. What distinctions come into question in the print, audiovisual, and online figurations of Wi-Fi? Some of them are the conventional ones—between private and public, between self and other, between inside and outside. While these distinctions remain equivocal, while they proliferate in variations and microdifferentiations, wirelessness as experience (in James's sense) marshals powers of affection that can both trigger further entanglements and attachments to products and services. The overall argument of this chapter has been that the contemporary ecology of wirelessness incorporates flows of desire and belief that provisionally stabilize in entanglements of locations, things, and figures.

Any disentangling of people and data in movement around wireless networks is complicated by the fact that many people are engaged in making sense of what other people are doing. The process of making sense of wirelessness belongs to wirelessness itself. As Massumi writes, for James, subject and object are "irreducibly temporal modes of relation of experience to itself"(108). Many if not all wireless projects or events attempt to alter not just the circulation of information (by removing wires), but our sense of people and data in movement. This recursive aspect of wirelessness is marked explicitly in the community and art-oriented wireless projects discussed earlier, such as Consume (Consume 2003) dating from late 2002 and ParkBenchTV (Gomes 2003) dating from 2003. The presence of Consume on a stand at a wireless networking trade show (WLAN 2003) promoting Wi-Fi to system administrators, IT directors, and architects complicates the distribution of ideas of consumers accessing information on the move. Recursive sortings of wirelessness inform the broader process that renders wireless products and services present. At one level, wireless networks are strongly associated with invisibility, with how to make network infrastructures less obvious and less closely constrained by existing building arrangements such as the need for wires to be put into walls, floors, or ceilings. But this exteriorization of wireless networks as infrastructures relies on fragmentary figures or ideas of people as wireless subjects. Any invisibility of information infrastructures, not just in terms of the invisibility of radio waves compared to copper wire or optical fiber, but in the form of increasingly febrile insertion of wireless technologies into existing devices, installations, and sites, relies on the formation of inner experience of subjects. Any drive toward making infrastructures invisible, as repressed content of the technological "unconscious" (Thrift 2004), however, vies with the need to quickly and immediately valorize Wi-Fi as a product that can be advertised, bought, and used in diverse locations. In this respect, wireless devices must also be sorted as outer. The variety of forms of externalization of meanings associated with Wi-Fi can be read, then, as oscillating constantly between the need to make visible, and yet allow something to become part of the background.[10]

A profusion of competing ideas of movement, space, access, and regulation make Wi-Fi difficult to categorize or even delineate as single object of experience. The lists in this chapter—locations, things, and figures—limn the growth and dynamics of Wi-Fi as cultural flow of beliefs and desires concerned with data and people in movement. Wirelessness is crosscut by vectors of commodification, competition between different telecommunication businesses, regions, and nations, challenges to infrastructural monopolies, and alterations in patterns of work and entertainment. As it becomes visible or culturally salient, it also changes. Different interpretations, and interpretations of other interpretations, circulate, not just in the form of statements but in images, modified objects, events, and somewhat novel social groupings. It is hard to hold the mixture of mundane repositionings (using a laptop in the lounge to do e-mail, etc.) and large-scale changes in movements of data (the growth of municipal networks in North America, Europe, and Southeast Asia) together. In contrast to the mobile phone (although Wi-Fi is very rapidly being integrated into mobile phones and many other devices), it cannot be said that Wi-Fi is "a device that provokes . . . strong longings and desires in people" (Townsend 2000, 98). Wi-Fi is not a device, but the trademark for an industry interested in continuous transition. Wireless networking, because it is not infrastructurally homogeneous or stable, offers a far more ambivalent case for analysis in some respects. However, as Wi-Fi gradually subsides into taken-for-granted network background, and as data percolates through a range of other wireless processes (WiMax, UWB, HSDPA, LTE, etc.), we should attend to the currents of feeling and the processes that render us affectable in wirelessness.

7 Overconnected Worlds: Development Projects as Verification for the Future

You can't lay cable. . . . It's difficult, expensive, and someone is going to pull it up out of the ground to sell it. (Greene 2008)

Intel, Cisco, Google, Microsoft, Huawei, and many large IT corporations invest in projects in Africa, South America, and Asia. Increasingly, these immensely wealthy enterprises make and market wireless equipment, products, and services in affluent Europe and North America based on images of what they do in less affluent parts of Africa or Asia. When Google announces its commitment to wireless networks in San Francisco (Mills 2005) *and* Abuja, Nigeria, as well as six other African cities (Paul 2006), or invests in a system of low earth orbit satellites over Africa "connecting the other 3 billion" (O3B Networks 2008) *as well as* wireless networks inhabiting interchannel television "white space" spectrum in the United States (Puzzangherra 2008), it implicitly associates cutting-edge technological innovation and development projects.[1] What work does this association do?

Almost 90 out every 100 people use the Internet in the Netherlands. Typically less than 1 in 100 use it in places like Chad, Somalia, Guinea, or Bangladesh (International Telecommunications Union 2009). People imagine, carry out, and resist globally oriented technical, political, ethical, and commercial actions through wireless networks in the name of this difference in network connectivity. Events, products, and projects based around building and managing wireless networks in developing countries have been bringing this difference to light. Wireless activists ("geek activism"), NGOs, businesses, and public-private partnerships engage in developing wireless networks in India, Southeast Asia, the Pacific Rim, and Africa. From the perspective of the North (Europe, North America, and parts of Australasia), wireless networking in and of the South both widens participation in global civil society and deepens entanglements with global neoliberal markets. The core tenet of many wireless development projects is that through wireless networks, others ("they") at the edges of the world may quickly become like "us" in the world cities and their environs. Nearly all of these projects seek to generate sensations of rapid change or transition: from remoteness to nearness, from poverty to prosperity, from exclusion to inclusion, from corruption to probity, from despotism to democracy, from rural to urban, and so on. As we will see, even as the very idea of development remains a somewhat fraught fantasy in many parts of the South, such a belief in wireless networks as world-changing impels wirelessness to overflow its existing location in cities, in houses, and in institutions in the North.

Drawing on global civil society literatures and recent network theory, this chapter analyzes how civil society and commercial wireless networks generate feelings of transnational or cosmopolitan relationality that allow wirelessness to create a sense of ongoing change and expansion. More philosophically it asks: How can we differentiate the senses of world at work in wirelessness as a contemporary determination of a world? It would, in many ways, be valuable to carry out empirical research closer to the rich and complex local practices involved in wireless development projects. I have not done that, relying instead on readings of accounts of wireless development that appear forthcoming releases, articles, field reports, and policy documents. The selection of different cases or projects is not meant to be exhaustive or very representative. Rather, my selection reflect events that attract the attention of newspapers or current affairs magazines, television shows, and websites. However, radical empiricist research differs from most newspapers, television, and websites in that it attempts to ascertain the "amount of either unity and plurality" in what counts as "the world."

Validate or Verify: Witnessing in Wireless Worlds

Wireless development projects done elsewhere (Rwanda, Nepal, Cambodia, etc.) always function as a spectacle to some degree. Viewed from a distance, the sight of wireless networks connecting people in arduous political, economic, or geographic settings to the Internet validates or verifies more proximate beliefs about the value of networks and being connected. In making wireless networks work "there," something about "here" is validated or verified. From a radical empiricist perspective, much hinges on the difference between "validate" and "verify." Validation confirms something that has already been mobilized or set in place. Verification, by contrast, relates to truth or verity, but pragmatically orients itself to consequences that affect action. Phillipe Pignarre and Isabelle Stengers (2005,

30) strikingly suggest that capitalism is not particularly pragmatic: "We would never say that capitalism is 'pragmatic.' Rather it is the anti-pragmatic par excellence, for the enterprise of systematic redefinition that it is associated with is not constrained by any verification, nor by any thought concerned about consequences."[2]

The key term here is *verification*. This is a contentious statement from the perspective of, say, an econometrician, but in terms of pragmatism, understood as a technique for the construction of ideas, capitalism does not verify, it only validates. Verification is an eminently practical procedure interested in consequences. It might entail experiments, tracking of consequences, and orientation toward further actions. Validation is anti-pragmatic in that it seeks to assert something and hold the assertion in place. It is resistant to overflow and overabundance.

What could be validated or verified in wireless development, especially if wireless network projects function as a spectacle? At first glance, it is often the technological know-how of telecommunications engineers or geek activists based in Cambridge, Massachusetts, or Berkeley, California, or Copenhagen, Denmark. Certain wireless projects, especially those with ambitions of connecting whole continents or regions to the Internet, perhaps more strongly enhance the prestige of their engineers or promoters than benefiting the places and people who may or may not make use of wireless networks. However, perhaps less often, locally developed expertise makes an appearance (for instance, in the extensive documentation of the process of setting up village networks around Dharamsala by AirJaldi (2007)). In any case, a spectacle is not intrinsically bad or alienating. Indeed, it may be that collectives inevitably cross-reference their own existence in the form of spectacle.[3] So, to treat wireless development projects as spectacle is not to diminish their significance. The anthropologist Marilyn Strathern describes a particular form of sculpture called *Malanggan* made in New Ireland, off the coast of Papua New Guinea, in which an intricate carving is produced in order to be seen for a few hours or days before being physically destroyed or discarded (sometimes by putting it into the hands of European traders). Witnesses to Malanggan play an important role: "Engaging the attention of the witness to the performance mirrors for the future what the performance itself encapsulates from the past in its own attention to an antecedent performance" (Strathern 2006, 23). The people who ostensibly benefit from wireless networks may or may not be witnesses to the performance of wireless development. (In fact, although I am in no position to confirm this empirically, critical accounts of ICT4D suggest that there are sometimes radical disconnects between the purposes and motivations of the people involved in setting up wireless networks and the people meant to use them.) Unraveling Strathern's quite complicated formulation a little, we might say this: if wireless networks for development "mirror for the future," they only do so by reference to past performances that come from "us," from "our" experiences and promises of communication. This puts a very different complexion on wireless networks. If they are performances that encapsulate something from antecedent performances, then they are not simply about a future in any linear sense. As spectators, we might see mirrored in them something that is not to our liking. For instance, Abdul Maliq Simone (2006, 151) describes the uptake of communication technologies in West Africa as distinctly at odds with the norms of open, transparent participation in global markets often predicated of information technologies: "Instead of the rapidly expanding increase in the volume of cell phone and Internet use being put to work as instruments to tracking transactions with greater scope, rapidity, and accountability, these tools are being engaged as means to intensify dissimulation."

We will see signs of these anomalies in the description of the situation in various places in Africa. For the moment, however, the main point is not what happens on the ground in West Africa or anywhere else. The question is how what the presence of past performances in the performance of development wirelessness shapes what we see.

Tensions between global and cosmopolitan relationality are intrinsic in wirelessness. Although William James is not usually read as a philosopher of globalization or politics, his work takes a direct and concerted interest in the mode of existence of "the world," and how worlds are known.[4] As I have argued in earlier chapters, wirelessness lies close to the most intimate operations and globally extensive logics of capital. In some of its structurings of what Nancy terms "being-with," wireless networks expose how these operations propagate on a global scale. The global expansion of wireless networks can be understood as a spacing out and connecting up of these structures. But they also encounter something that cannot be entirely subsumed or captured by processes of validation and valuation. We might say, they encounter a world.

According to radical empiricism, a world coheres in provisional and shifting ways. James (1911, 131) writes: "From the point of view of these partial systems, the world hangs together from next to next in a variety of ways, so that when you are off of one thing you can always be on to something else, without every dropping out of your world."

James argues that there is neither pregiven unity or plurality to the world. His account of different scales of connectedness implies the need to constantly demonstrate or perform new kinds, levels, or degrees of partial connection. Without connection, or "hanging together," one could drop out of "your world." Similarly, Nancy's (2007, 43) account of what constitutes a world is relevant since it addresses edges, and movement to edges: "The stance of a world is the experience it makes of itself. Experience (the *experiri)* consists in traversing to the end: a world is traversed from one edge to the other, and nothing else" (Nancy 2007, 43).

Again, the scope of this formulation is philosophically ambitious, yet it can be read pragmatically. Taking a "stance" on a world scale means going to the edge of the world. Yet this stance only comes from a world making itself in movement. The edges and ends, and traversals to the edge of a world, come to matter in development projects not only as ways of taking a stance on a world, but as a world. A world is its traversed edges. To speak of *a* world implies that there might be other worlds. Nancy and James's positions converge on the question of the oneness or unity of a world. James (1911) considers the question of "world unity" at length in *Some Problems of Philosophy.* He asks:

What connections may be perceived concretely or in point of fact, among the parts of the collection abstractly designated as our "world"?

There are innumerable modes of union among its parts, some obtaining on a larger, some on a smaller scale. (p.126)

As always, the approach James takes to the question of a world is to verify what differences mark experiences of a world. His stance on the world depends, like Nancy's, on "traversing to the end." However, he suspends any decision as to whether the world is one or not: it is in some ways, and in other ways not. The composition of the world is singular-plural. As James writes, "The amount of either unity or plurality is in short only a matter for observation to ascertain and write down, in statements which will have to be complicated, in spite of every effort to be precise" (pp. 133–134).

Becoming pragmatic about the world itself means experimenting with the very idea of a world, experimenting with the construction of different ideas of a world or worlds, and verifying their effects. Hence the "observations" written down this chapter draw out a series of wireless development projects across Africa and India in their connections to North America and Europe. I argue that development wireless catalyzes the promissory value of wirelessness in the North.

Wirelessness, as a contemporary determination of the world, is not simply another manifestation of globalization. Globalization attempts to configure the world as a unity. In his discussion of globalization, Nancy (2007, 43) distinguishes globe and world, globalization, and world formation: "The world is thus outside representation, outside its representation and of a world of representation, and this is how, no doubt, one reaches the most contemporary determination of the world."

What does Nancy mean by saying that "the world . . . is outside representation"? Much of this chapter portrays the forms of awareness and participation running through wireless networking for development projects as exposures of "being-with" at the edges of the representation of the world as global. From a radical empiricist standpoint, the crucial questions in framing wireless development between world and globe, verification and validation, are: (1) How do these exposures of world edges connect to a becoming-world? (2) How is an "us" constituted in this exposure that mixes validation and verification?

Like Nancy's account of being-with, radical empiricism does not take a world as immediately given. The singular-pluralism of experience as James accounts for it does not necessarily take place in a world as such. Rather, a world comes into being for James (1911, 126) by following and making connections: "What connections may be perceived concretely or in point of fact, among the parts of the collection abstractly designated as our 'world'?"

In earlier discussions of radical network empiricism, I have suggested that the challenge of thinking wirelessness partly arises from the proliferation of conjunctive relations that occur around wireless networks. These relations impinge on gestures, feelings, habits, and sensations in ways that are largely subrepresentative, or that are organized in substitutional series (for instance, in the idea of the wireless city or the wireless community). Conjunctive relations undergird the identifiable and visible forms of subjectivity, power or place that people inhabit as they play the roles of academic, activist, commuter, engineer, manager, or student. For instance, these relations are coupled with the intensive movements of signal processing that mingle and redirect everyday movements into and away from processes of commodification and service provision.

Wirelessness has within it an implicit awareness that these everyday relations are distributed very differently in different places: some places on earth have many more kinds of network services and products on hand than others. This difference is usually highly significant. Or at least, it can be made to signify something amid the fluxes of telecommunication network change that have been occurring for the last few decades. Amid an abundance of networked media, there is something particularly captivating about places that lie a long way from well-developed infrastructures of telecommunications and transport to network connectivity and its associated forms of liberalized markets. Such locations include deserts, oceans, and mountains, as well as war zones and slums. Wireless development projects make these differences become visible, and perform them as significant. As I will argue, making transnational and global differences significant has been a constitutive dynamic in wirelessness, a major component in rendering of change in network connections as especially rapid, effective, cheap, and indeed, potentially universal. This instrumental signification of development wirelessness runs alongside another component of wirelessness that is less easy, yet arguably vital to also identify. In augmenting a radical empiricist sense of the world, I read Nancy's account of being-with, and of the experience of being-with, against what happens at the edge of "our" experience, in places that function as peripheries or perimeters.

Valuably Remote Zones: Timbuktu

From 2002 onward, a series of articles, press releases, and conferences announced the deployment of wireless networks in the global South. Dozens, perhaps hundreds of wireless development projects, with a wide range of actors involved, were animated by the idea of Wi-Fi supplanting, bypassing, or bridging the paucity of communication infrastructure. They ranged from relatively localized ICT4D projects such as Wi-Fi-equipped motorbikes in Cambodia (Jhai Foundation 2003) to projects with global ambitions such as the $100 One Laptop Per Child (OLPC) project with its ad hoc wireless connectivity. The proponents of wireless development range from GeekCorps to Cisco and Intel, from the United Nations to the International Telecommunications Union (ITU). The politics of ICT4D and the nature of strategic investments made by transnational corporations such as Intel, Cisco, Google, and Motorola in wireless networks in "bottom-of-the-pyramid" places such as Laos, Nepal, Mali, Sri Lanka, or Nigeria call for careful consideration. The visibility of Wi-Fi development projects in Western mass media and online media has been an important contribution to the expansion of wireless networking in North America, Europe, and Australasia. The connectivity of wireless networks is very often equated with a rapid transition from being "off the map" to being full economic actors in global markets. While the photographs, diagrams, and written descriptions of exploits, difficulties, obstacles, and inventions that appear in the news and online concerning wireless development may appear and disappear quickly, they help experiences of development wirelessness to "get themselves reproduced" in many versions (Strathern 2006, 19). Most recently, these developments have focused on mobile phones in Africa and India (Zachary 2008).

While these different sites are widely scattered in terms of geography, they form a *technological zone* in the sense proposed by Andrew Barry (2001). As he defines them, *technological zones of circulation* "are spaces formed when technical devices, practices, artefacts and experimental materials are made more or less comparable or connectable. They therefore link together different sites of scientific and technical practice. Such zones take different forms. The points of access to the zones may be more or less clearly marked, with more or less well-defined and functioning gateways" (pp. 202–203).

Berkeley *and* Kigali, or Taipei *and* Phnom Penh, can function as points of access to a shared technological zone of wirelessness. Wireless equipment offers a particular version of making comparable or making connectable. These places undergo incorporation in zones of wirelessness when development projects link them and make them connectable to the Internet. In the zone, a poorly connected periphery gains access to a center, but a center animates and enlivens its own sense of value by connecting to the periphery. Wireless development projects function very much as a perimeter for a technological zone of wirelessness. Their peripheral location supplies something to the much more intensely connected zone of wirelessness in cities and regions of North America, Europe, and East Asia. If this connection between wirelessness and peripheral development is not coincidental but generative, then what animates their relations?

In one sense, wireless development projects offer a particularly pure form of wirelessness. It particularly addresses the absence of communication infrastructure. For instance, the much publicized One Laptop Per Child (OLPC) association produces rugged, low-cost, laptop computers called XO for children. OLPC, a highly prominent wireless development project, minimizes assumptions about any existing infrastructure. It promises a wireless infrastructure—a mesh network—in the absence of "external infrastructure":

Mesh Network

Right out of the box, the XO laptop's antenna ears can sense other neighboring XO laptops and connect to them, creating an instantaneous "mesh" network for communal sharing and collaborating. This is important because many XO laptops will be deployed in places where there is little or no external telecommunication infrastructure. (OLPC 2008)

Sociality takes the form of "communal sharing and collaborating." In the context of development wirelessness, as elsewhere, social practices of collaboration are regarded as directly transformable into economic value (via the education of children as future participants in a global marketplace of skills). The version of sharing and collaborating imagined here depends on network infrastructure.

However, the development of a wireless network in the absence of "no external telecommunication infrastructure" always encounters traces of other connections. Again, there is something akin to the Malanggan here. The burning of New Ireland Malanggan described by Strathern has an important function in mirroring for the future antecedent performances. Its destruction engages the attention of the witness. However, thousands of Malanggan actually exist in museum collections across Australia, Europe, and North America (Sykes 2004). Similarly, there is no pure zone of wireless development. For instance, in the case of the city of Timbuktu, in Mali, Africa, a place that signifies remoteness from the North or the West, new wireless development projects need to be mindful in planning networks that a number of previous projects have already equipped Mali with Internet and mobile phone connectivity. There are many antecedent performances of connectivity or communication associated with Timbuktu. The very image of Timbuktu as remote or "off the map" has attracted many development and commercial wireless projects. In some respects, and not by any means for all its inhabitants, Timbuktu has changed almost in step with the network society. A tourist guidebook to Timbuktu comments: "What's extraordinary, though, is that while physically Timbuktu is almost as isolated as it was in the 1970s, in terms of wireless communications, it's completely on the map. Mobile phone coverage has just reached the town. There's European satellite TV in the hotels and giant dishes poking out of mud hut compounds" (Trillo 2005). Development projects are often quite aware of antecedent performances. In a case study of a project in Timbuktu, one of the authors of *Wireless Networks in the Developing World,* Ian Howard, describes antecedent performances of connectivity:

Timbuktu is remote, though having a world renowned name. Being a symbol of remoteness, many projects have wanted to "stake a flag" in the sands of this desert city. Thus, there are a number of information and communications technologies (ICT) activities in the area. At last count there were 8 satellite connections into Timbuktu, most of which service special interests except for the two carriers,

SOTELMA and Ikatel. . . . Relative to other remote cities in the country, Timbuktu has a fair number of trained IT staff, three existing telecentres, plus the newly installed telecentre at the radio station. The city is to some degree over saturated with Internet, precluding any private, commercial interests from being sustainable. (Flickenger et al. 2006, 213)

Rather than wireless development taking place in absence of infrastructure, it must often negotiate various kinds of antecedent performances.

There are different ways in which it can do this, pragmatically and somewhat antipragmatically. Some of the authors of *Wireless Networking in the Developing World* set out to construct a wireless connection between a small rural radio station and a "telecenter" located in the court of the mayor's offices. "Telecenter" refers to a place where various people can access computers and computer networks. The telecenter movement in Africa and South Asia is large and organized, with extensive involvement of NGOs, governments, educational institutions, development agencies, and other groups.[5] While myriad personal digital gadgets and wide saturation with networking equipment cluster at the central technological zone of wirelessness, the telecenter represents a key point of access to the edge of the networked technological zone. Many wireless development projects pivot around varieties of telecenters. A telecenter typically has a connection to the Internet. In many parts of Africa, the costs of connecting to the Internet are high, since it usually passes via commercially run geostationary satellites rather than fiber-optic or copper cables. Hence the Internet connection often depends on external funders, or it might be shared between a telecenter and the local offices of a Western NGO. Already, wireless development projects have to make affiliations with other development projects.

Many reports on wireless development projects describe the technical and infrastructural design in great detail. "At this site, the team decided to implement a model which has been called the *parasitic wireless model*" (Flickenger et al. 2006, 212). The key feature of this model is its extreme simplicity. It consists of a bridge, a wireless access point that connects two existing networks. The technical details of their implementation might seem irrelevant. However, in the context of wireless development projects, hardware and software arrangement take on a special significance. They are not merely technical details, but a core part of the performance of connecting or making connectable. In a sense, the design and installation of hardware and software comprise the performance we are asked to witness. It is the principal way in which the often impossible tensions between politics, economics, culture, and geography—tensions that might destroy the project—can be managed. While the ongoing life of a wireless network in Timbuktu is quite uncertain (and difficult to even investigate), its ongoing existence can only be taken into account in certain ways. Even when sustainability is a key concern (for instance, as it is in much of the community informatics and telecenter movement literature), this is presented in terms of technical design of power supplies, and so on. In the Timbuktu case, the developers describe the installation quite carefully:

In this installation the client site is only 1 km away directly by line of sight. Two modified Linksys access points, flashed with OpenWRT and set to bridge mode, were installed. One was installed on the wall of the telecentre, and the other was installed 5 meters up the radio station's mast. The only configuration parameters required on both devices were the ssid and the channel. Simple 14 dBi panel antennas (from http://hyperlinktech.com/) were used. (Flickenger et al. 2006, 213)

This description is quite heavily laden with detail intended for other people working in wireless development. There is a "client," the "rural radio station," which is in line of sight of the telecenter at the mayor's office. The construction of a 5-meter-high mast, something whose ongoing existence would be very much taken for granted in the global North, can be a significant technical challenge in its own right in certain settings. While setting up a wireless network is not too hard, connection to the Internet via satellite remains very expensive. However, the next points of the description invoke a whole bundle of features that cannot be reduced to the topography or everyday life in Timbuktu. Compressed in this brief description are the algorithmic folds of 802.11 wireless digital signal processing with all the complications of chip economies, the version of the Linux operating system known as OpenWRT that can be installed ("flashed") on certain brands of wireless access points such as Linksys, and the antennae purchased from a company in Florida that specializes in "disaster recovery and rapid deployment communication products" (HyperlinkTech 2007). Most of these components—the chips, the access points, the operating system, the antennae—were not created to be used in developing countries. Rather they stem from the practices of connecting and modifying wireless networking components that we have seen in earlier chapters. It is as if these practices now search for validation in development work.

The technical details described in the project both grapple with the antecedent performances of connecting or communicating in Timbuktu, but at the same time idealize a technical expansion of wireless networking. Looking back on the project, the developers conclude, "As it stands,

Internet access is still an expensive undertaking in Timbuktu. Local politics and competing subsidized initiatives are underway, but this simple solution has proven to be an ideal use case" (Flickenger et al. 2006, 215).

On the one hand, "local politics and competing subsidized initiatives" (presumably other development projects) display awareness of antecedent performances. On the other hand, the "simple solution" is an "ideal use case." The formulation might sound strange. "Use case" is a term from 1980s–1990s software engineering referring to modeling techniques that "capture requirements" for a software system.[6] It lies upstream of anything actually being put into the world. But here, the term, perhaps invoked casually, means that the wireless network they set up in Timbuktu becomes a mirror for the future expansion of the zone of wirelessness. Tension between future expansion and antecedents runs through many wireless projects. What the "ideal use case" involves here cannot be reduced to the hardware the Timbuktu project effectively put in place in the telecenter and the rural radio station. However, there is a strong sense in which putting that hardware together in a "simple solution" does something important at the edges of the Internet. It augments the sensation of being in a zone of wirelessness that can expand.

A "Worldwide Lust for Technology": Validating Events, People, Products, and Projects

It is not surprising that much development wirelessness takes the form of projects in specific remote locations, and in particular, in places that are somewhat visible or significant by virtue of their remoteness (for instance, Timbuktu in Mali, Africa). These projects link wireless networks with future markets for products as well as with labor. To participate in the globally connected markets, and to experience the rewards of that participation, people in the global South will need to be connected to transparent, well-governed systems of transaction. The Internet is uniquely suited to support those connections. Wireless networks in particular can overcome the infrastructural deficiencies that afflict these newly emerging markets, and they can cheaply circumvent the telecommunications bottlenecks maintained by unreformed state monopolies (especially in sub-Saharan Africa). Above all, wirelessness promises that this can be done quickly in many places at once, whether through mobile phones or Wi-Fi networks.

Where would we locate wirelessness in relation to development projects and the politics of development more generally? Development projects over the last few decades have been heavily criticized for being colonizing, homogenizing, masculinist, extending state power or neoliberal economic reorderings, as well as being depoliticizing. In an overview of the field of development studies, the academic field that both analyzes and designs development projects, Stuart Corbridge (2007, 180) writes: "It is right that the concept(s) and practice(s) of development are rendered problematic. We also need to understand that the origins of development studies were closely linked to the beginnings of a Cold War between the First and Second Worlds, and that the broader development business is often beholden to geopolitics."

According to Corbridge's analysis, the very notion of development is beset by impossibilities of various kinds. Development tends to depoliticize change by technologizing it, thereby muting questions of power, inequality, or exclusion. Critics of development argue that post–World War II and Cold War development policies have produced the Third World in various ways, and at the same time, through organizations such as the World Bank and technical development projects, attempted to ameliorate various humanitarian, economic, and political crises triggered by the very making of the Third World. Such attempts often end up extending state bureaucratic power or legitimating power elites, as well as institutionalizing a development industry that unwittingly or helplessly participates in the production of political, humanitarian, economic, and environmental crises. Above all, development suffers from developmentalism, an attitude that holds that "'they' can be quickly made like 'us'" (Corbridge 2007, 189) in the First World.

Certainly a belief in or hope for one world, at least in relation to Internet connectivity, often accompanies wireless development projects. However, in some respects, wireless networks seem to take some of these problems into account. They bring with them an aura of small-scale, grassroots activity, independent from states or corporations. They valorize local connections (for instance, the sharing and collaborating done through a wireless mesh of XO laptops). However, validation and verification, antipragmatism and pragmatism, edge uncertainly around each other in wireless development. The development projects occupy ambivalent political and economic positions. They range across civil society actors, public-private partnerships, and "global development alliances."

Wireless development projects reflect various forms and scales. Sometimes they appear at major, highly visible media events that publicize the possibilities for whole countries or regions of rapid change or "leapfrogging" associated with wirelessness. These include large congresses and conferences sponsored by the United Nations or the International

Telecommunications Union (ITU) such as the World Summit on Information Societies (WSIS), which endorse wireless networks as bridges for the digital divide. More recently, mobile telephones have also received similar attention in the context of development. Articles have appeared in *The Economist* and the *New York Times* (Dabba2008, Zachary, 2008). Such events are very much staged as global, for the sake of the institutions and organizations that see themselves as global. Wireless networks appear as a way of integrating lives into global networks. At a conference on technology in 2003, the UN Secretary-General at the time, Kofi Annan stated: "It is precisely in places where no infrastructure exists that wi-fi can be particularly effective helping countries to leapfrog generations of telecommunications technology and infrastructure and empower their people" (United Nations 2003).

The Annan announcement is also deeply suffused with ambivalence since "empower their people" could either mean political power, presumably in the form of democratic enfranchisement, or it could mean economic power, presumably through fuller participation in globalized neoliberal markets. It is perhaps the latter value that most often prevails. At the same conference at which Annan spoke, it was reported that Pat Gelsinger, chief technology officer of chip maker Intel, claimed that "[Intel] was getting huge amounts of interest about wireless networks from many poorer nations. 'This reflects a worldwide lust for technology,' he said. 'We see millions of people with the potential to become wi-fi users'" (BBC 2003). In this case, an enterprise logic embodied in Intel can be only be expressed as a global desire—a "worldwide lust for technology."

There are also smaller, widely scattered conferences such as ICTD (ICTD 2007) in Bangalore, "Wireless Roadshow" (wire.less.dk 2007) in Tanzania, "Wireless Africa Workshop" (FMFI 2008) in Cairo and Pretoria, or "AirJaldi" (Airjaldi 2007) in Dharamsala, India. These are sponsored by a mixture of private and public organizations. At such events, specific wireless projects are often discussed. They include significant efforts to train or transfer technical knowledge of wireless networking to the global South in order to overcome the uneven distribution of technical know-how. For instance, each year since 1998, workshops on wireless networks for developing countries have been run at the Abdus Salam International Institute for Theoretical Physics, Trieste, Italy. The result of these workshops, at least publicly, is what the *International Herald Tribune* headlines as "Wireless: The Tin-Can Antenna Offers a Boon for Third World" (Povoledo 2006). Significant exchanges of expertise occur at these events, and the websites for these events often vouch for the presence of a mixture of entrepreneurs, NGO activists, academics, technicians, and engineers.

However, the most visible wireless development projects plan on a large scale, say a billion people or so. The much discussed $100 or XO laptop, or Meraki "Free the Net" systems, would be examples of these more ambitious projects. For instance, Meraki Networks Inc., a wireless Internet equipment and service management business, was founded in 2006 by MIT graduate students. It hopes to add another billion people to the existing 1.5 billion Internet users. Based in San Francisco and funded by venture capital from Sequoia Capital and Google, Meraki sells wireless hardware adapted for outdoor use and wireless network management services:

Meraki Outdoor

Three devices in one tough package.

Includes Meraki's Hosted Services

The Meraki Outdoor is a weather-proof and UV-protected wireless access point, gateway, and repeater, designed to boost your network's range out of doors. Built to withstand everything from hackers to extreme temperatures, water, sand, and dust. The Outdoor contains a high-powered radio and long-range antenna for a robust, worry-free plug-and-play deployment on building exteriors. And every Meraki device delivers the entire Meraki experience, with instant access to Meraki's unique Hosted Services. (Meraki Networks 2008)

As we have seen in earlier chapters, such products come and go rapidly. However, Meraki tries to prolong the life of its products by "delivering the entire Meraki experience": software to allow organizations to manage their wireless networks as well as built-in software that automatically associates access points with other Meraki devices in a wireless mesh. These organizations might be schools, trailer parks, airports, or housing associations. More importantly, they might be telecommunications companies that wish to deploy wireless networks as cheaply as possible in India or Brazil. Ultimately, the Meraki experience might also include people from among the billions who do not use or have any chance of using wireless networks. In a 2007 article in *Scientific American,* this possibility is discussed as if it explained everything else. The article is titled "Meraki's Guerilla Wi-Fi to Put a Billion More People Online" and reports the ambitions of one of Meraki's founders, Sanjit Biswas: "Biswas's hypothesis is that empowering individuals to create their own networks, and perhaps even profit from them, makes it inevitable that grassroots efforts will spring up to bring

wireless Internet access to areas where it is currently unavailable or pro-
hibitively expensive" (in Mims 2007).

"Empowering individuals to create their own networks' here means
supplying them with hardware that can be managed by Meraki using mesh
networking techniques the founders Sanjit Biswas and John Bicket
researched when they were PhD students. In fact, Meraki's most visible
wireless network is found in San Francisco, where "Free the Net" allowed
people to either associate their own wireless access point with the Meraki
network, or to buy an access point such as the Meraki Outdoor or Meraki
Indoor and add it to the network (Meraki Networks 2008). Over 100,000
people belonged to the network in early 2008. "Free the Net" in San
Francisco replaces muni-wireless projects that Google and EarthLink tried
to set up there in the years 2005–2007 (see the preceding chapter). It also
functions as a testbed for plans to deploy similar similar mesh networks
in other places such as Harlem, New York, as well as to "the next billion
people" in over 125 countries (Meraki Networks 2008). Meraki's "Free the
Net" encapsulates the logic of many wireless development products writ
large. Access to the Internet is provided at low cost or in some cases for
free to individuals in the North, but at the same time, the very notion of
wireless development promises wireless connection as a way to concatenate separate worlds (First and Third). Meraki and many similar projects
ride on the assumption that supplying network connections to people
living in the poverty of disconnection or separation will also promote the
project organizers' expertise in deploying or managing wireless networking
in high-tech urban settings in San Francisco, Copenhagen, or Boston.

High-Speed Wireless Development

How would Meraki Outdoors fare in Africa today? This can only be a matter
to ascertain empirically. The ambition or hope for the "next billion people"
(or in the case of the recently announced 03B "Other Three Billion"
project, the "other three billion" (O3B Networks 2008)) motivates experiments with wireless equipment in remote, disconnected, or impoverished
locales. Why remoteness as the primary criterion? One answer to this question lies in the very notion of the "global." Anna Lowenhaupt Tsing (2005,
76) writes that "since the 1990s, every ambitious world-making project has
wanted to show itself able to forge new scales." Scales can be made, proposed, practiced, and evaluated in different ways. Sometimes new scales
are forged very literally. Development projects have competed to extend
the range of networks well beyond the few hundred meters envisaged by

the IEEE standards. Attempts to make Wi-Fi work on a different scale have been numerous, and they continue. They often involve modifying antennae and towers to work over several hundred kilometers (e.g., the world distance record for a wireless link is held by Berkeley's Technology and Infrastructure for Emerging Regions project (TIER 2008) in Venezuela); redesigning power supplies (solar, batteries, etc.) to allow wireless infrastructure to work unattended in rural or remote locations; developing software control of signal propagation (e.g., the Intel Research Group in San Francisco has worked on software control of direction and now sells a product, the Rural Connectivity Platform (Greene 2008), so that networks maintain links when even wind, children, or animals move the towers); mixing wireless networks and other means of transport (e.g., United Villages Inc. based in Cambridge, Massachusetts, uses wireless buses to collect and deliver e-mail in villages in Orissa, India (United Villages Inc. 2008)) to intermittently connect places to communication networks. In the context of development work, spatial scale, topology (for instance, Meraki Mesh's ability to automatically form networks), and time become the focus of much technical work. Some of these projects have truly global ambitions. Eric Brewer, a computer scientist who works at Berkeley, envisages TIER's Wireless Long Distance (WiLD) projects in India and Africa as a way to "give people in rural areas hope and mitigate the urban migration" (Merrit 2008).

But wireless development projects forge scale in other, less obviously world-spanning ways. By far the most common form of development wirelessness takes the form of "projects" involving at most several hundred people. These projects are usually located in village settings, and they connect schools, hospitals, small businesses, health clinics, medical centers, telecenters, and cybercafés to the Internet (for e-mail, Web, and VoIP). Occasionally in Africa they can be found in larger urban settings such as Accra or Kinshasa. At least insofar as they are visible on the Internet, these projects rely heavily on donations of time, equipment, or expertise from the North. Because such projects always have a prospective aspect, it is hard to track exactly what happens to them over time. The smallness of the project does not necessarily mean that they are of little importance.

The difficulty of seeing on what scale wireless development moves also stems from the problem of actually locating a development "project." As discussed above, the viability of many development projects, and the very idea of a development "project" itself has been critically contested. But there is something more to this instability than local contingencies or large-scale economic, political, or social challenges. When Kofi Annan decreed that wireless networking was the way forward for developing countries to participate in the information revolution (BBC 2003), he may or may not have envisaged the "infrastructural inversion" (Bowker 1994) occurring today in wireless development. Rather than wireless development bringing "our world" to "their world," it would be possible to argue that in certain respects wirelessness validates itself worldwide.

The proliferation of high-speed commercial WiMax and other wireless technology trials and networks in Africa, Southeast Asia, and less affluent parts of Eastern Europe would be one example of this validation. Although technologically akin to Wi-Fi, WiMax takes on very different forms and infrastructural roles. Because it has much greater range, it can readily support infrastructures that either compete with cellular phone and fixed-line telecommunications or form hybrid infrastructures with them. While WiMax is a product of the computer and network industries rather than the telecommunications industry, unlike Wi-Fi, it sometimes use expensive licensed segments of the radio spectrum. When market competition becomes fierce or regulation too restrictive, the telecommunications industry can use places such as South Asia, Africa, and the Middle East to vet wireless technologies in trimmed down, lightly regulated forms. Just as in the clinical trials of pharmaceuticals in South Asia and East Africa (Rajan 2003), wireless development becomes a leading edge of an innovation thrust, rather than an afterthought to it.

From 2001 onward, advanced forms of wireless connectivity can be seen appearing in fairly unexpected locations alongside attempts to verify much more localized, low-tech wireless networks. For instance, we would expect Taipei to have advanced wireless networking given its location and its high-tech economy. Taipei's advanced wireless network (as we saw in chapter 2), and indeed Meraki's "Free the Net" in San Francisco, like many municipal networking projects, are "mesh" networks (Herman 2005). Mesh networks (or simply "wireless mesh") differ from the star or the hub-spoke arrangements of many wireless and wired infrastructures. By contrast, the wireless development projects usually present much simpler connectivity. Wireless networks are found in and around African cities such as Kigali (Rwanda), Abuja or Lagos (Nigeria), Addis Ababa (Ethopia), or Porto Novo (Benin). Here Wi-Fi networks are said to be growing rapidly. The networks in these places, we would expect, lack the concentrated network density of First World global-city wirelessness with its wireless meshes. Yet if we focus just on Kigali, Rwanda, connectivity is a complicated issue. Rwanda is "East Africa's number one ICT nation," according to the UN Confererence on Trade and Development (Rwanda News Agency/Agence Rwandaise d'Information 2007). Kigali, its main city, presents very differently sited wireless projects laid over each other. On the one hand, ICT development projects such as First Mile Solutions' "Hybrid Real-Time, Store-and-Forward WiFi Mesh in Kigali Rwanda" (First Mile Solutions 2007) invoke the notion of "mesh," and thereby suggest that the same idea of "filling gaps" is being applied. It provides "realtime access . . . for several sites" (First Mile Solutions 2007), although "store-and-forward" implies that other sites are only connected to the network intermittently. On the other hand, commercial enterprises such as Altech Stream Rwanda, a subsidiary of a South African telecommunications company, was licensed to blanket Kigali with Wi-Fi and WiMax (a newer, faster, longer-range wireless networking technology that, unlike Wi-Fi, can use licensed spectrum) by the end of 2007. There is a marked contrast between the First Mile Solutions in its efforts to connect schools and Altech Stream Rwanda with their focus on "business and consumer segments, . . . government and foreign institutions in the educational, diplomatic agencies and embassy offices" (Creamer Media 2007). The large connectivity disparity between the First Mile Solutions network and the Altech Stream Rwanda network suggests that in these places, wireless networks participate in much more speculative forms of connecting and comparing connections. The fragility of existing infrastructures and lack of market competition support experiments in proliferating connectivity.

The Overconnected State and Telecommunication Anarchy

Kigali ends up as the target of modest ICT4D projects and cutting-edge commercial wireless networking at the same time. Rwanda's situation represents something more general about wireless development, especially in Africa. Wireless development, we could say, "overconnects" the edges of the world. It subjects them to more than one kind of connection, to both high-speed, wide-coverage networks and to relatively sparse, selective network connections. The overlay of very different wireless connectivities in one place can result in an unstable state of affairs. Rapid changes in wireless networking, telecommunications, and the Internet are occurring in many parts of Africa. However, the presence of wireless networks in African cities sometimes appears as a threat to economic development. Wireless networks installed by development projects or by private individuals clash directly with state regulation of telecommunications and media. For instance, in the West African state of Benin, a 2007 government announcement attempted to corral a proliferation of different networks, including wireless networks:

Benin's telecoms: total anarchy, draconian measures announced.

A study commissioned by the Government of Benin of the country's telecom sector has made a "bitter report" on the anarchy which prevails there, provoking the government to announce a series of drastic measures, *Ouestafnews* has learned from an official source.

The study reveals that a total of 47 operators on 50 inspections, carrying out "all or part of their activities in violation of regulations," deprive the State of Benin of "significant receipts" and cause "a powerful financial haemorrhage damaging to the State of Benin." . . .

The state, according to the communique is going to also suspend all "orders authorising provision of telecommunications services such as VoIP, local loop radio, wifi, wimax, adsl, pre-paid cards with the exception of internet service providers and cybercafe operators who conduct their activities legally." (Ouestaf 2007)[7]

Although Wi-Fi and WiMax are merely mentioned among many other networking technologies (such as ADSL, local radio loops, cybercafés, etc.), Benin's response is not unusual, or asymptomatic. The government of Benin is not alone in reacting harshly toward the "anarchic state" that has emerged as a result of people becoming involved in setting up and running wireless networks. There are many other examples in Africa and the Middle East where wireless networks are heavily regulated or even prohibited. Rwanda's situation is, in fact, atypical. In neighboring countries, wireless networks have a much more precarious position. In Gabon, a US$65,000 license fee is imposed. In 2007, Niger still prohibited both voice and images on Wi-Fi networks. Loss of telecommunications revenue through unlicensed wireless networks can affect the fiscal stability of developing countries. For different reasons, the situation in the Middle East, with the exception of Iraq and its commercial-military infrastructures, is even more restrictive. (See Stichting Open Spectrum 2007 for comparisons of licensing and regulation of Wi-Fi networks in various countries.)

It would be a mistake to regard this ambivalence on the part of states concerning wireless networks as confined to the global South. At the same time that development agencies were promoting wireless networks as a path to economic development, U.S. federal government agencies were starting to treat wireless networks as a risk to national security. In 2002 one could read in *Wired* magazine that "'(an attack) could bring down the network of this country very quickly. Once you're on the network, it doesn't matter where you got in,' said Daniel Devasirvatham, who headed the Homeland Security task force for the Wireless Communications Association International trade association" (Boutin 2002). In response, "[The U.S. Department of] Homeland Security is putting people in place who will be in a position to say, 'If you're going to get broken into . . . we're going to start regulating,' said Cable and Wireless security architect Shannon Myers in a panel dubbed 'Homeland Security vs. Wi-Fi'" (Boutin 2002). Wirelessness here lies close to the potential telecommunications anarchy of Benin, it seems (and this is borne out in more recent events: the Mumbai bombings in India in 2008 involved the use of unsecured wireless networks to send e-mails and the use of VoIP services to maintain contact between the attackers (Cox 2009)). Hence, a year or so later, U.S. Deputy Secretary of Defense Paul Wolfowitz, soon to move to the World Bank, signed a directive, *Directive number 8100.2 Use of Commercial Wireless Devices, Services, and Technologies in the Department of Defense (DoD) Global Information Grid (GIG)* (Wolfowitz 2004). Given that the U.S. military already accesses and controls many parts of the usable radio-frequency electromagnetic spectrum, it is ironic that its most global communication system was dogged in places such as Iraq by a profuse array of commercial wireless devices dangling off its networks. Wolfowitz's directive responded pointedly to the growth of unregulated and unmanaged wireless networks in the hands of U.S. troops in Iraq. Media and communication devices had a very ambivalent status in that war. While the conduct of the war relied deeply on global communication, communication networks also allowed troops to communicate with each other and with their friends and families. This led to an unplanned, unmanaged expansion of the global network-centric warfare platform, GIG, through consumer electronic devices such as laptops, digital cameras, and mobile phones. Unable to enforce a complete decoupling of the consumer gadgets from the military GIG, the 8100.2 directive made a compromise.[8] "Commercial wireless devices," such as Wi-Fi, can be attached to the GIG if everything is encrypted. The most core platform of contemporary networkcentric warfare, the GIG, can henceforth include consumer electronics gadgets and perhaps rely on devices such as Wi-Fi routers and network cards? Henceforth, GIG is an "enterprise service (ES)," and GIG is referred to as GIG ES in Department of Defense documents. A notion of the enterprise runs deep in the contemporary state.

Even if wireless networks constantly evade regulation by states, the rapid transition to wirelessly networked cities can be too rapid and overconnected for some purposes. Put differently, wireless connectivity, especially in the difficult-to-regulate form of Wi-Fi, need not contribute directly to economic development. Rather than promoting participation in global networked economies (as is the hope for Rwanda), wireless networks can afford something different. Wireless networks are not always in the service of increasingly fluid integration into global systems of trade and commerce (tracking transactions with greater scope, rapidity, and accountability). They can form part of shifting, ephemeral, networks of activity and aspiration that have little regulatory visibility. This can be as simple as making friends in other places through Internet chat. It can sometimes offer radically different models of entrepreneurship that only become visible in the form of notorious Internet scams and fraud. Field research that compares mobile phones and the Internet in Ghana, for instance, backs Simone's general argument in relation to wireless networks in African cities. Anthropologists Don Slater and Janet Kwami (2005) suggest that Internet use in Accra, Ghana "is [mainly] a utopic familialization of relationships with strangers in order to secure the structure of obligation and reciprocity that is normative for local and embedded relationships." That is, rather than the Internet or wireless networks serving as a way to get the "next billion people," or the "other three billion people," online as consumers, producers, workers, or spectators, they can also serve as ways of helping people cope with the increased uncertainties and perturbations associated with global connectivity. This key dimension of Internet use in much of Africa bypasses the framing of connectivity in ICT4D wireless networks and the rising experimental investments in high-speed, cutting-edge wireless networks as testbeds for Northern markets. In both rich and poor countries, in both the hypercommunicative e-City of Taipei and the fragmented urban–rural African peripheries of cities such as Kigali where First Mile Solutions overlaps with Altech Stream Rwanda, Wi-Fi seemingly instantiates the ideal of connectivity. This connectivity is meant in both cases to be free of the legacies, frustrations, limits, and expense of other, earlier infrastructures (such as roads, sanitation, railways, etc.).

It may be that in African cities this ideal appears in its barest, stripped-back form. Development wireless projects, in their aspirations to connect Africa, part of Asia, the Caribbean, Pacific Islands, or Latin America into the global economy, do not register the limited gains that connection to the Internet offer to many people, especially when "urban Africa experiences even more inexplicable rhythms of sudden accumulation and loss" (Simone 2006, 136). In the contrast between cities such as Taipei and Kigali, what is striking is not so much the common ideal of wireless mesh as the epitome of connectivity. That seems to be shared to a greater or lesser extent. Rather, different kinds of overconnectedness appear. From this perspective, the Wi-Fi or Wi-Max network draped over the streets of downtown Kigali is the figure of problematic and unstable change. Simone (2006, 155) suggests that "Urban Africans have continuously been involved in remaking their cities in ways that enable them to use the shifting agglomeration of different bodies, stories, and spaces as a means of livelihood. In this way, urban life has a sense of home-grown virtuality. This virtuality is being combined with the virtual environment opened up by ICTs."

Rather than ICT4D being the means of developing countries participating more fully in globalization from a disadvantaged position, they form part of a flexible system that treats splintered or absent infrastructures in peripheries as catalysts for product development and rapid innovation. Ironically, in Taipei, the comprehensive municipal mesh, a highly visible symbol of national economic competitiveness, struggles to compete with proliferating minor wireless networks and hotspots. Simultaneously, in certain African cities, the proliferation of the same minor wireless networks appears as a threat to state monopolies of telecommunications. The severity of state regulation of Wi-Fi networks in many African and Middle Eastern countries points, at least in part, to the existence of forms of lateral connectedness that seek to evade and circumvent the transparency and visibility of the municipal mesh networks.

Collectives of Connection: We Are Hailed by Development Projects

Wirelessness is a "contemporary determination of the world," to use Nancy's term (Nancy 2007, 43). Other contemporary determinations of the world include "climate change," "war on terror," or "human rights." None of these determinations exist in isolation, and indeed wirelessness is cross-hatched into all of them. The comparison between what happens in Taipei and Kigali is not farfetched, for wireless networks for development and wireless networks for urban mobility, entertainment, and productivity both rely on conjunctive relations, the forms of being-together, coiled within wirelessness itself. In a sense, wirelessness expresses its limits in wireless development. There, in the unstable forms of conduct and practice discussed above, lie "specific and verifiable connections" (James 1911, 124) that make worlds hang-together or fall apart.

A mosaic of contradictory tendencies in war zones, in cities and villages in Africa, the Middle East, and Southeast Asia, affect wireless development.

As we have seen on several occasions, radical antiempiricism could be said to exist alongside radical network empiricism. Frequently, wireless development projects do little more than validate a belief in expanding networks. They make, as Wendy Chun (2006, 143) writes, "it possible to believe once more in liberal and consumer equality." The process of validation applies to many wireless projects, especially the public-private partnerships sponsored by Intel, Cisco, Google, and Microsoft, but also the liberatory projects of Meraki or TIER: they seek to validate oneness of the world, and to establish a general equivalence through that. They engage in an "enterprise of systematic definition" (Debaise 2005, 106) that submits everything it encounters to the ideal of isotropic network connectivity regardless of consequences. In some wireless development projects, indifference to verification or care for effects can be seen writ large. What is at stake in these projects is rather a promissory speculation, a construction of a discursive mirror in which North American, European, or increasingly East Asian futures can see themselves. However, this figuring of the edges of the world is problematic and unstable. We have seen how state forms of capture slip when it comes to wireless networks. Although cellular phone and fixed cable infrastructures remain under state control, wireless networks are harder to hold onto.

Some wireless development events, even in their gaps, their shortfalls, and their limits, can be seen as a struggle to verify something without general equivalence of power, value, or property. The effervescent quality of experience that James's radical empiricism seeks to addresses recurs in wireless development in West Africa or southern India, in the efflorescing varieties of virtuality that do not conform to the open, transparent, accountable mechanisms of wealth creation promoted by the World Bank or USAID. There is a certain pragmatic experimenting and verifying of the possibility of connection that goes on. The technical accounts of wirelessness in less developed countries in their attention to how a piece of PVC tubing can be converted into an antproof wireless access-point enclosure, or how a wireless network can operate given the uneven, intermittent power supply in Nigeria, can be seen as involved in the invention of forms of transmission.

In his account of how capitalist productions of value reverse into an "egalitarian, singular, and common significance" (Nancy 2007, 49), Nancy writes of "transformations of the evaluation of value" (p. 48). He states that "our task today is nothing less than the task of creating a form or a symbolization of the world" (p. 53). The symbolization of the world takes the form of worldhood. In some ways, this lies at odds with James's radical empiricism, in which the world could take on no determinate form except for specific, verifiable, practical ends. However, the form of symbolization Nancy (2007, 53) advocates is practical and verifiable: "One could say that worldhood is the *symbolization* of the world, the way in which the world symbolizes in itself with itself, in which it articulates itself by making a circulation of meaning possible without reference to another world."

A positive reading of wireless development could be framed in terms of worldhood, or as creating a form of the world that does not refer to another world. Worldhood has no reason or ground or signification as such (economic development, consumption, human rights, belonging, locality, humanity, etc.). Its "meaning" is circulation, the "possibility of transmission from one place to another, from the one who sends to the one who receives" (p. 52). Indeed, as Nancy (2000, 28, 3) writes, "There is no other meaning than the meaning of circulation." As we saw earlier, Nancy proposes that a bare relationality is exposed in communication networks that says something about what "we" might be today. From this perspective, wireless projects in the streets of Kigali or outlying villages of Cambodia are the figure of "'us' as web or network, an *us* that is reticulated and spread out, with its extension for an essence and its spacing for a structure" (p. 28). The experiment of experience "consists in traversing to the end," as Nancy (2007, 43) writes.

Verification and *validation*: these two terms describe different trajectories and organizations of wireless development. Wireless development is not simply an extension of something that has occurred from "here" outward, to the edges of the world. It is not an added or peripheral dimension of wirelessness. Rather, it displays highly varied forms, speeds, and temporalities that attest to different forces in play. In some ways, I have argued that validation tends to be primary in many of the projects, products, and partnerships associated with wireless development. Ultimately, validation asserts the primacy of sameness through practices of planning, implementation, and measurement. Verification, however, understood here in a pragmatist sense, engages with the plurality of things, beliefs, and connections amid abundance, instability, and change.

8   Live, Forced, Momentous Options and Belief in Wirelessness

We bathe, according to James, in an atmosphere traversed by large spiritual currents. (Bergson 1934, 243)[1]

As James was writing, a large-scale wireless world was gelling in a shape initially staked out around massive electromechanical radio stations drawing hundreds of kilowatts owned by the nation-states or a few corporations in North America and Europe (Marconi, Bell, Siemens, etc.) (Aitken 1985). These stations served to reinforce and project naval and mercantile power across seas. The monumental radio towers of the early twentieth century radiated signals in a world economy centered on London and New York. They competed commercially in an economy already connected by a global network of underwater cables (Mattelart 1996, 2000). In the early twentieth century, spectrum was allocated in a similar way to colonial territory. Whoever occupied a portion of spectrum came to own it. In the late twentieth century, the cable and satellite transmission capacity of the world is somewhat greater and radio spectrum is much more densely populated, especially in cities. The global capacity of the Internet is mainly limited by the submarine cables that link different continents. That capacity is reported at 7.1 terabytes per second (Telegeography Research 2007). By comparison, the overall capacity of the totality of Wi-Fi connections based on estimated sales of 802.11b/g/n chips (ABIResearch 2007) lies somewhere around 20,000 terabytes per second—if they were all connected to each other, which they are not. (This estimate is ridiculous since there is no way at present to manage such an infrastructure.) Unlike submarine cables, which have been laid over the course of 150 years, wireless networks uncoiled in the space of seven or eight years (1999–2007). Wi-Fi networks are just one somewhat fragile and tenuous incarnation of wirelessness, especially in comparison to the much more prominent and widespread cellular phones, with their many masts and antenna towers. However, the cable versus Wi-Fi comparison shows that we are witnessing a significant moment of expansion in network culture, at least as measured in potential connections and network bandwidth.

This raw capacity is not the result of telecommunications companies speculatively draping optical fiber in thick bundles on seafloors and city streets, as they did in the 1990s in expectation of massive media change. Nor does this capacity arrive in the form of a public utility or infrastructure, although certain municipal wireless networks have been motivated by something like a reanimated hope of a public utility. Wireless network capacity is a strangely disaggregated and fluxing topology of small-scale links that offer partial connection and semi-open forms of association and aggregation—local areas, communities, campuses, cities, and so forth. On the one hand, there can never, it seems, be enough network capacity. As Geert Lovink (2003, 370) writes, it seems that "never enough Internet capacity can be provided to the velocity-hungry online masses." On the other hand, there is vastly more capacity than is actually used. For instance, only 29 percent of the trans-Atlantic fiber-optic cable capacity is said to be "lit" in normal use (Guardian 2008). "Dark fiber" and its unrealized potential certainly has a wireless phantom associated with it. If wireless network capacities are tens of thousands of times greater than cable capacities, then surely the potential of wireless devices could to an even greater extent be said to be "yet to be realized" (Lovink 2003, 376). While most of this bandwith operates over very short distances compared to submarine optical fiber cable, the problem of "dark fiber" has been often been explained in terms of what happens in the "last mile." In the last mile, in the interval between the network node and the network periphery, many different things can happen. The smooth celerity of photonic crystal optical fiber running along the seabed and cross-country suddenly frays into the messy, accident-prone, constantly crosscut, dug-up streets of cities and towns. The streets are already packed with cables, tunnels, sewers, drains, vehicles, and people. Hence in the "last mile," we could say, as Lovink (2003, 376) does, that "the net is developing in possibly conflicting directions."

What happens to wireless network capacity today is not easy to ascertain or visualize. By 2009, data had replaced voice as the major form of growth in mobile telecommunications businesses (Gabriel 2009). Mobile telecommunications service operators are themselves uncertain about what is happening in the growth of wireless data in their networks. They "build out" network infrastructure (such as 3G "mobile broadband") on the assumption that large quantities of data will need to move quickly. ("Build out" is the term used by wireless operators, but it derives from urban planning, where it designates an estimate of the amount and location of potential development for an area. In other words, "build out" is not the process of installing infrastructure, but a process of deciding what development potential an area offers.) However, their planning seems to be falling awry of much of the way that wirelessness works. Enormous industry investment has focused on increasing bandwidth. Belief in bandwidth motivates national government policies, investment by mobile network operators, as well as the projects and policies of international bodies such as the International Telecommunication Union (ITU) and IEEE (Institute of Electrical and Electronics Engineers). Actually, although large chunks of data (video, music, images, files, etc.) are moving through networks more, the major unanticipated form of growth in mobile data appears to consist of the flood of small messages generated by many different kinds of applications rapidly or periodically polling for updates (Ghadialy 2009). E-mail and location-based services make up only a small part of network traffic ("bandwidth"), yet consume large amounts of signaling capacity ("airtime").[2] Different applications have very different impacts on the wireless network. "Airtime" ("the radio network access time to deliver applications" (Ghadialy 2009)) and "bandwidth" usage display very disparate perspectives on what is happening in wireless networks. "Data is not data in a wireless network," explains Mike Schabel, general manager of Alcatel-Lucent's network analysis product, 9900 Network Guardian (Alcatel-Lucent 2008). If that is the case, and analyses of contemporary network traffic suggest it is, then belief in continually growing network bandwidth as the defining feature of better networks begins to seem crude. The billions of devices populating wireless networks seem to both heighten such a belief and make it less tenable. In some ways, wireless mobile devices (such as iPhone, etc.) highlight the irreducibility of wirelessness to the networks even as, in other ways, they obscure it.

The Froth of Transition and Wireless Tendencies

The fact of wirelessness comes into existence on different temporal and spatial scales, some of which move more quickly than others. Wirelessness effervesces or foams (Sloterdijk 2004) rather than flows. Can one simply believe that today wireless devices make networks better? This book traces interconnected tendencies affecting wirelessness between 1999 and 2009. The tendencies are not entirely specific to wireless devices and network

services, or to wirelessness as a mode of contemporary experience, nor are
they reducible to a belief in better networks through more bandwidth and
more wireless connections:

• *Conjunctive envelopment and relational manifolds*   Contemporary digital
signal-processing-based communication environments in Wi-Fi, 3G,
Bluetooth, WiMax, LTE, UWB, CDMA2000, and so on, enfold people and
things in a patchwork of signals, overlays, and connectivities. This patchwork weaves together intensive differences at the level of fluxes of information and radio signals with conjunctive relations generated by bodies and
things in movement. While information follows many paths through
contemporary networks, the specificity of conjunctive envelopment goes
deeper. It relates to a plethora of spatial, urban, and infrastructural features
while fostering trends toward network universality and network expansion. Sometimes the complicated mathematics of digital signal processing
(or other algorithmically regulated processes) is regarded as the very essence
of abstraction, dematerialization, or decontextualization. I have argued, by
contrast, that the mathematical intricacies work with basic conjunctive
relations, relations that shape the diverse flows of experience. Conjunctive
envelopes shroud perception and movement. They begin to microtune
subrepresentative relations of proximity and distance, before and after. A
priori conditions in their bare minimal forms undergo redesign, modification, and regulation. The many wireless devices that have appeared, and
that seemingly inevitably will appear, are folds or wrinkles in the conjunctive envelopment.

• *Consumer individuation partitioned across infrastructural being-with*   The
promissory horizon and the experience of wirelessness is riddled with
inconsistencies and tensions. There is no pure wirelessness, only transitions that bifurcate repeatedly between consumption and the bare, exteriorized form of conjunctive relationality, being-with. On one side stands
the slightly depressing euphoria of personalized, monadic individualized
mobility, dazzled by ideas of wireless cities, wireless communities, wireless
work, and wireless worlds. Tens of thousands of terabytes on the move
furnish individuals with personalized attachments to highly processed
media and online services: "Technology makes it possible to believe once
more in liberal and consumer equality," as Chun (2006, 143) writes. On
the other side, a swarm of marginal and seemingly peripheral socialities
and practices generate mutated network-oriented infrastructures constructed around antenna modifications, substitute firmware, node databases, and so on. So many frustrations and disappointments attend

wirelessness, but they also necessitate new habits and skills in maintenance, configuration, and modification of the compartments ascribed to
devices and infrastructures. In many domains, we encounter hesitant,
tenuous trajectories of networked action that do not originate in human
actors but come from entanglements with signals, locations, and others.
As networked devices sink deeper into the conduct of a life, as forms-of-life
become device-intensive, desires to set and realign boundaries on devices
operate with greater intensity. In these bifurcations, new kinds of crimes,
ethical dilemmas, policies, controls over access and security, and new
senses of proximity and distance arise.

• *Recursive overflows of networked mediation* Wireless experiments, variants, federations, alliances, partnerships, collaborations, and hybridization
of gadgets, devices, infrastructures, and transactions proliferate in tension
with the figure and practices of networks. The twentieth century saw a
steady stream of such overflows in telegraphy, telephony, radio and television broadcasting, and satellite communications. Wireless media have
begun to act recursively to produce new networked arrangements and
ensembles of people, sometimes close to the Internet (as we saw in many
mapping and database projects that use the Internet to construct maps and
diagrams of how to use the Internet), but sometimes a long way from any
easily recognizable figure of the network (for instance, in the proliferation
of wireless logistics tags). Wirelessness as experience of transition accommodates an enormous variety of intermediary, imitative, and overflowing
fluxes. Recursive intermediaries—databases, roaming plans, wireless-
enabled gadgets, and so forth—appear as conjunctive relations undergo
verbalization and exteriorization. Hype in such environments is a still
mutable and unstable reality. It attaches and detaches, entangles and dis-
entangles, and readily unravels differences between inside and outside.
Suspended between interior and exterior, between thinking and thing,
affectional experiences permit these recursive overflows to propagate and
multiply.

• *World making as concatenation on the edges of experience* A key trait of
James's radical empiricism is that any "whatness" of experience glides on
a platform of conjunctive relations. Conjunctive relations support a sense
of continuous transition, and indeed, they allow one experience to pass
into another: "There is no other nature, no other whatness than this
absence of break and this sense of continuity in that most intimate of all
conjunctive relations, the passing of one experience into another when
they belong to the same self" (James 1996, 50). At the same time, these
conjunctive relations are increasingly organized, marshaled, and federated

in ways that bind experience into fixed formations. If globalization is a systematic enterprise of redefining aggregates that burgeoned in conjunction with communication and transport networks, consumer electronics and network gadgets can be seen as personal experiential components in that enterprise. Wirelessness, however, could be said to juxtapose awareness of the crumbling edges of world-global differences with consumer electronics and gadgets. New actors such as wireless development start-ups often seek to couple a sense of global social responsibility with experiments in remote wireless infrastructures. The extensive array of wireless network development projects coincides with the aggressive capitalization of advanced wireless networks in peripheral zones where market conditions are radically different. Infrastructural inversions occur in which the most advanced communication technologies appear in the global South before they gain purchase in Europe or North America.

A reader in a synoptic mood might construct a table in which each of these pathways or tendencies could be correlated with indices of value, space, materiality, identity, time, and power. A broader grasp of these tendencies would benefit from an exploration of their interaction to produce a belief in wireless networks.

Belief in Wirelessness

In the *Will to Believe,* James (1960, 4) concludes that belief is required when we face a choice between two or more "live, forced, momentous hypotheses." His account of the nonideological production of belief begins by making some philosophical distinctions. James defines a "hypothesis" as any notion open to belief. "Live," in this context, is used in an electrical sense ("just as electricians speak of live and dead wires"). In general, it is only really worth exploring "live hypotheses" (James 1960, 2). A dead hypothesis refuses to "scintillate with any credibility" (p. 2) and would only be of relevance to "absolute duffers" or to one "who has no interest whatever" (p. 21). All of this is consistent with radical empiricism's understanding of ideas as particularly rapid or advantageous pathways. The degree of liveness of a hypothesis correlates with its propensity to lead to decisive action: "The maximum of liveness in an hypothesis means willingness to act irrevocably" (p. 3). In turn, a choice is "forced" when there is no way not to choose: "go out or stay in" is a forced option since you have to choose one path, and by not choosing, have a choice imposed on you. James pays "momentous" little attention, simply opposing "momentous" to "trivial," although, as usual, it would be possible to argue that even this distinction is not banal. All of these distinctions—live versus dead, forced versus unforced, momentous versus trivial—help James argue for the necessity of "passional decisions," and a "will to believe," first of all in a religious sense, but also in a sense that would affect any thought. The essay sanely concludes that belief is intellectually indefensible but necessary.

James's essay might be profitably read in the context of a collective passion or "worldwide lust" for wireless networked telecommunications. In a sense, this book has investigated contemporary belief in networks. In the case of wireless networks, we have seen the degrees of liveness, forcing, and momentousness of something quite mundane, which oscillated between live and dead, momentous and trivial in the last decade. The book has tracked the evolution of hypotheses and beliefs associated with wireless connections on display in post-dot-com network cultures. Wireless devices are sometimes framed by beliefs in technological inevitability that are hard to question. James calls any decision between hypotheses an "option" (p. 3). To decide is to opt. What kind of option is associated with wirelessness? In what way does it involve "passional decisions"? Many accounts of the Internet, information or network society, new media, and digital cultures address changes in global politics, media, civil society, publics, audiences, representation, identity, power, law, space, time, and embodiment. Much of this work shows how people—users, subjects of the net, consumers, citizens, or netizens—are transfixed by the Internet, and the vistas of network connectivity. However, such accounts often skip over the underlying dynamics that generate belief in digital connectivity as the solution to a problem, as an option offered to belief. In this respect,a radical empiricist treatment of experience for wirelessness offers something different. If it is increasingly hard to imagine a world without wireless networks, then we need to understand how "live hypotheses" concerning relationality, sociality, space, and materiality come into being. Faced with live, forced, momentous hypotheses, belief, in the pragmatic sense of a potential to act, is required since "reason," including those varieties embodied in cultural and social studies of technology, cannot decide. Belief is a form of attachment animated by the superabundance and redundancy that radical empiricism identifies in all experience. In radical empiricist terms, belief or willing-to-act arises in situations that cannot jettison live, nontrivial, and forced options. If we follow James's account of belief as willing to act, there is no room for agnosticism in relation to any of those aspects of experience from which belief arises: liveness, momentousness, and forcing. A politics and ethics of wirelessness are not based on reason, but on passional decisions.

Liveness in the Last Few Years: "Comparatively Effervescent"?

The liveness of wirelessness is temporary, and certainly it is partial. Wirelessness might only be a live issue for a short period of time. What does it mean to think about a half dozen years of change via a specific technology that operates on a relatively small scale (usually less than a kilometer)? What is generalizable from the concrete minutiae of wireless networks in the last decade, compared to the decades-long growth of the Internet and the century or so of information and cybernetic culture? Events can be situated on different scales, but the time scale occupied by wirelessness concerns near futures and recent pasts. In understanding how near futures emerge, the problem is how to decide what counts as a near future or recent past.

We could handle the problem of liveness via an "anthropology of the contemporary" (Rabinow 2003, 55). In Paul Rabinow's (2003, 55) view, anthropological inquiry is a field-based experimentation in relocating thinking in relation to "distinct temporal scales." In *Anthropos Today*, Rabinow sets out these scales (drawing heavily on Foucault's account of eventalization). He defines philosophical anthropology as a specific practice that takes place in an encounter with a complex temporal conjuncture. One of the things that makes an anthropology of the contemporary hard to do is the "distinct temporal scales" that crosscut it. On the broadest of these scales, problematizations emerge. Problematizations are large-scale events that reshape what counts as thinkable. Rabinow (2003, 46–47) shows how Foucault distances himself from the history of ideas, from the problematic in Althusser's sense, and from the "unthought" in Heidegger. Problematizations are events that deploy thinking in "the elaboration of a given situation into a question" (p. 47). They do not entail solving that question, or making it coherent, but making it into an "object of thought" (p. 49). Thinking, wherever it occurs (in the sciences, art, popular culture, politics, and technologies), is central to problematization: problematization names the process whereby actions or practice become a matter of thought, not as a matter of evaluation, or as a matter of conscious intention, but adrift between interpretation and practice. Problematizations then are large-scale historical events that link an ensemble of discursive and nondiscursive practices, of ways of making sense, and of ways of doing things. In Rabinow and Foucault's terms, we might say that "information" and "network," and through them, much network-attached technology today, are the locus of a problematization. In many discourses, institutions, media, and in many lived modalities, differences between true and false, real and fiction play out in terms of concepts, practices, and forms of information and network. The networked informatization of the world flows out of many practices that become objects of thought. Ontological differences are increasingly linked to computability, but what is computable cannot be directly decided, and has to be renegotiated constantly (e.g., in relation to security or total information awareness).

For Rabinow, again following Foucault, these epoch-making problematizations contrast with *dispositif* or apparatus: "The apparatus is a specific response to a historical problem" (p. 54). An apparatus is, says Foucault, "strategies of relations of forces supporting, and supported by, types of knowledge" (p. 53). Technologies and techniques mostly reside at the level of *dispositif* or apparatus. The apparatus has shorter-range stability: "Apparatuses are the forms composed of heterogeneous elements that have been stabilized and set to work in multiple domains" (p. 55). For instance, throughout information technologies of various kinds, one constantly encounters the same techniques of calculating, approximation, sorting, ordering, and copying techniques. These techniques, developed in response to specific problems, gradually turn into general techniques carried into other domains, sometimes with unexpected results. We could think of the generalization of certain compression techniques like the fast Fourier transform derived from telecommunications across many audio and visual media, as well as information infrastructures such as wireless networks, as an example of apparatuses set to work in multiple domains. Some of the most animated developments in recent information and network cultures result from the generalization of the strategies of sorting and ordering developed in the context of twentieth-century science, commerce, or war into popular culture. Apparatuses define strategies, objectifications, and subjectifications, and install general logics across many domains. So for instance, the many gendered aspects of contemporary subjectivity connected to technologies could be seen as apparatuses, as generalized stable forms that propagate across domains.

Rabinow offers a final vital contrast in temporal scales and modes: the assemblage. (Indeed, he claims to principally be interested in assemblages rather than problematizations or apparatuses.) What needs to be analyzed in relation to technology is the matrix out of which assemblages emerge. Assemblages are "comparatively effervescent, disappearing in years or decades rather than centuries" (p. 56). They emerge as events in contingent relation to long-duration problematizations, and sometimes soon disaggregate. Intermittently, an apparatus emerges from them. This contrast between the time scales of problematizations, apparatuses, and assemblages frames the anthropology of near futures and recent pasts. As Rabinow says, his anthropology "seeks to identify emergent assemblages . . . [and] to identify conjunctures between and among these diverse objects, and between and among their temporalities and their functionalities" (p. 56). While little in Rabinow's discussion treats conjunctive relations explicitly (at least to the degree that James does), his emphasis on assemblage as *conjunctures* of temporalities and functionalities of diverse objects does resonate with radical empiricism's attention to conjunction.

In the "comparatively effervescent" years between 2001 and 2009, wireless infrastructures introduced new conjunctures into diverse networks. In these years, wireless networks attracted intense imagining, figuration, political contest, inventive practices of location and modification, thinking about infrastructure, and new modalities of service and product. Wirelessness could then be seen as an assemblage emerging in response to the network problematization. An assemblage-oriented wireless anthropology would look for things that appear to have the capacity to differ from themselves, that are difficult to locate, that seem to harbor incoherences or incompatibilities triggering divergences. In general, these could be divergences in relation to social functions of gender, class, or race, in sense of self and other. They could be divergences in relation to logics of production, circulation, or consumption, or divergences in relation to systems of state power, publics, and the organization of capital.

Approached as an assemblage, the relatively small scale and short life cycles of wireless devices, standards, and infrastructures are more a strength than a weakness. The small and rapid changes associated with them generate conjunctures among objects, temporalities, and functionalities. A central argument of this book has been that wireless connections embody an infectious set of relations—conjunctive relations—that make spaces and times both cohesive and mutable. On this point, radical empiricism diverges from an anthropology of the contemporary. Conjunctive relations hold assemblages together without the substantial unities or divisions associated with relations of identity and opposition, and without the overarching temporal scales of problematization, apparatus, and assemblage. In settings where everything slides, James's radical empiricism suspends distinctions such as subject versus object, represented versus representation, thinking versus thing. It searches for the consequences that flow from them. As these distinctions undergo pragmatic processes of verification, their foundations dissolve. James (1996a, 233) ends up arguing that all we can say is that things and thinking come from the same stuff of experience in general. His account of truth differs from any representation. As Bergson (1934, 246) writes, for James, "Reality flows, we flow with it, and we call any affirmation 'true' that steers us through this moving reality, that gives a hold on it, and places us in the best conditions to act."[3]

Experience is the name given to something that can overflows any verification, any exploration and or experiment. For James (1996a, 36–37), even thinking is largely another name for breathing: "The stream of thinking (which I emphatically recognize as a phenomenon) is only a careless name for what, when scrutinized, reveals itself to consist chiefly of the stream of my breathing." For any ontological footing in matter, ideas, subjects, or objects, he substituted a provisional processing of tendencies, traits, expectations, trajectories, and feelings unraveling in things or thoughts taking shape. If experience encompasses mixtures of objects and subjects, individuals and collectives, thinking and things, the work of radical empiricism consists in attending to conjunctive relations, to relations such as withness, nearness, and betweenness that are normally sidelined. These small-scale, often short-lived relations allow assemblages to spatially and temporally cohere. These relations permit something like a flow or stream of experience to occur. Their liveliness in making subjectivity, matter, sensing, remembering, and acting cannot be underestimated.

A radical empiricist account of wirelessness unearths some of the fractalizing processes through which network connections become viable on different scales, scales that often geographically exist on the order of a few kilometers or miles at most, and perhaps for a few years at most. The many devices, projects, practices, images, standards, policies, programs, and organizations described in the preceding chapters spin out from tendencies in transition with different rates of realization. In wirelessness, the expansive spatialities and temporalities of networks or cyberspace are tested against many boundaries, thresholds, and forms of openness and enclosure. The wireless networks discussed in these pages all in their own ways allow new claims to be made about how networks extend and information propagates (for instance, in relation to development), but also quickly put these claims to the test at various levels. While they are not ultimately consistent or unified, all of these tendencies comprise wirelessness. The hundreds of different variations, the billions of chips, the countless individual struggles, frustrations, and satisfactions of maintenance, repair, connection, and access attest to the liveness of wirelessness as an option. These different tendencies have and make some appeal to belief, in a value of wirelessness.

Momentous: "Temporalize in Resisting the Present"

The wireless world is growing at an astonishing rate. (AlcatelLucentCorp 2008)

The second core dynamic that generates beliefs, according to James, is something momentous in a situation. What is the momentous aspect of wirelessness? If anything, it resides in perceptions that it means growth and change. The problem of making sense of change resonates strongly throughout the radical empiricist approach to wirelesssness I tested in the previous chapters. As James (1996a, 161) writes, "'Change taking place' is a unique content of experience, one of those 'conjunctive' objects that radical empiricism seeks so earnestly to rehabilitate and preserve." Wirelessness is, effectively, a concept of experience modeled on contemporary things in constant change. The method of following conjunctive relations by crisscrossing between thinking and things does something more than affirm assemblages as intersections of different things. It transposes us into matrices of experimentation out of which forms of experience emerge.

On this point, the development of radical network empiricism as a method converges with one of Henri Bergson's rules of intuition as philosophical method: "State problems and solve them in terms of time rather than space," as Gilles Deleuze (1988, 31) puts it.[4] Like James's radical empiricism, which starts by asking, "Are thought and things as heterogeneous as commonly said?" (James 1996a, 28), Bergson's philosophical method of intuition constantly seeks to move to the edges of experience. As Bergson (1934, 181) writes, "We call . . . intuition the *sympathy* through which one transports oneself into the interior of an object in order to coincide with what is unique and consequently inexpressible in it." There are close alignments and perhaps even reciprocal borrowings between Bergson's method of intuition and James's radical empiricism.[5] Stating and solving a problem in terms of time means stating it in terms of change, and more specifically, change that comes from differences in kind. Posing the problem of the growth wirelessness in terms of time in turn means analyzing changes in terms of differences in kind. The name for the power of differing in kind in Bergson's philosophy is well known: *duration*. Duration is change in kind. Space is the totality of differences of degree. Change in spatial terms is augmentation and diminution (of proportion, of figure, of interval, etc.). Things spatially differ in degree. Durationally, however, things differ in kind. Things differ from themselves through alteration. So wireless networks differ greatly in degree, and they project many differences of degree. Many of the network enterprises, projects, plans, ideas, topologies, maps, and images associated with Wi-Fi indeed aim to control differences in degree: in distance, in scale, in speed, capacity, or number. Yet they also change themselves (in their associativity, in their modes of collective existence); and this change is irreducibly part of how they exist. In his account of Bergson's philosophical method, Deleuze (1988, 32) writes: "This alteration, which is one with the essence or the substance of a thing, is what we grasp when we conceive of it in terms of Duration." Alteration is the "substance" of the thing. Radical empiricism also has a great affinity with this perspective.[6]

Differences in kind undercut any easy mixture of differences in degree. The differences in kind I have in mind here are *tendencies*, not essences. Yet if experience is always composite, how could we ever distinguish and divide the different tendencies, the differences in kind? Is not the degree of experiential mixing too great to ever separate tendencies? We have already seen that, for James (1996a, 94), experience is composite all the way down: "Far back as we go, the flux, both as a whole and in its parts, is that of things conjunct and separated." Like the Bergsonian method of intuition, a radical empiricist approach would begin from those aspects of experience between which "*there could not* be a difference in kind" (Deleuze 1988, 25). There are terms in wirelessness in which there "could not be a difference in kind." For instance, the opposition between wireless and wired might be regarded as a difference in kind that fundamentally structures any experience of wirelessness. The contemporary world is divided by wired and wireless connections. Are these different kinds? On the basis of the analysis of wireless signal processing or the work done to make even local wireless connections, we could quite easily say that the only difference between wired and wireless is in degree of movement. The difference between wireless and wired networks lies in where or to what degree wires are entangled with each other. The wireless chip is only a somewhat more compressed tangle of wiring than a wired network interface. This difference in kind is actually a difference in degree between more or less highly introjected spaces of conjunctive relations.

What occupies the interval opened up by differences of degree between wired and unwired? The "astonishing rate" of growth of the "wireless world"? For James (and Bergson), affectivity and belief come to fill the interval. (Remember that for James, *affect* refers to that dimension of experience that is not yet sorted, aggregated, or fused into patterns of utility, communication, and identity.) Affectivity, recollections as linking between instants, and memory in the form of bodily habit: these all come into play in wirelessness, as we have seen. Wirelessness infuses movements and perceptions with affectivity concerning connection, communication, location, and other dimensions. The experience comprises tendencies that cannot be represented, at least not within the turn or fold of "our" experience. To even sense the force of tendencies means adjusting ourselves to flows of alterity. This adjustment is difficult: "The acts of intuition . . . have to be multiplied," writes Deleuze (1988, 27). The analyses offered in previous chapters of this book have attempted to multiply such acts in different domains and scales of wirelessness. In the convoluted interval between network connections and airtime, new sensitivities, new orientations, new connections, and new orderings of action embody themselves. Wirelessness itself, as a set of experimental practices, performs acts of intuition in various ways. Attempts to map, to reconfigure, to federate, to extend, and so on, all entail protointuitive movements that feed both pragmatic experiment and antipragmatic enterprise. The radical network empiricist has to multiply acts of intuition that follow the diverging lines or tendencies up to the turn that subtends lines of experience as a domain of "our experience." One bizarre and dizzying final move in the method of intuition must be added. To access the specificity of wirelessness as a singular admixture, the method of intuition must follow different lines or tendencies to a point at which they converge beyond "our" experience, a point that gives the precise reason for a specific experience.

What would be the result of an effort to be radically empiricist in relation to contemporary networks? We detect differences in degree running underneath what appear to be differences in kind (wired versus wireless). We find differences in kind—that is, durations or tendencies—that embody themselves in differences of degree. Finally, we go beyond the turn that delimits our experience to find where the tendencies converge again. The aim is to generate a highly specific concept, or as Bergson (1934, 23) puts it, "a concept modeled on the thing itself, and which, in this sense, is no broader than what it must account for."[7] Much of this book has been an exercise in trying to accompany alterations in the fabric of networked communications. The constantly unraveling collage of examples, materials, cases, and events has a durational aspect. It can be seen as the attempt to adapt to alterability, to the substance of change, and to the change in things. We have seen the distinctly effervescent temporality of wireless assemblages. That effervescence stems from the primacy of tendencies over the essence in things. The temporality of assemblages varies constantly, sometimes crystallizing as apparatus, sometimes fluxing into problems and problematizations.

Forced Belief

Beneath the local transfer there is always a conveyance of another nature. (Deleuze 1988, 47)

How does "live, . . . momentous" belief in wirelessness end up also being "forced," as James puts it? The live, momentous aspect of wirelessness plays out in the wireless cities, the node databases, the intensively wrought architecture of signal-processing chips, or the world-changing ambitions of the development wireless projects. In wirelessness, acts of intuition multiply as many different people seek to differentiate and articulate different tendencies in the expansion of networks. The liveness—in James's sense—of wirelessness comes from that aspect of it that configures space ordered to our needs, actions, and interests. Wirelessness interests us to the extent that it diminishes some kinds of separations and augments others in relation to networks. However, forcing seems to be another matter entirely.

Today, just when wireless service providers realize that they do not know exactly how wireless data moves through the networks or what it means for their planned infrastructures, Wi-Fi has also become a control system for factories, sensor networks, logistics management, and theme parks: "Disney is using an industrial-strength 802.11 wireless network to power its Toy Story Mania! ride, which opened last year at both Disney's California Adventure in Anaheim, Calif., and its Hollywood Studios in Orlando" (Duffy 2009). The network allows communication between the cars carrying riders and large 3D screens showing characters from the *Toy Story* films. It is a key component of the illusion of the ride since it allows coordination between movements of the cars, the riders, and what appears on the 3D screens. Here the wireless network has become a control system for the temporary suspension of disbelief. The force of wirelessness increases as networks sink into the intervals between movement and perception.

After it has sunk into that interval, there is no other option apart from wirelessness. A key challenge for radical empiricist researchers of wirelessness (or technologies of life itself, etc.) would be to find the "specific means" or singularity of that sinking. The "force" of wirelessness comes from the specific means or singularity of its situation. The approach I have taken here is to show how pragmatic attempts to work with the conjunctive relations generated at the interfaces between air and wires, between airtime and network time, generate a power to intervene. Wireless networking is a domain pervaded by commercial realization and antipragmatic tendencies. It perhaps comes into existence as a commercial-public project in the same way that "open-source" software is a commercial-public phenomenon. (We could also include the free software projects of the late 1990s, or sound artists and musicians working with electronics and software.) The powers to intervene in wireless networks are perhaps few and far between. Whatever specificity of means they establish is likely to be quickly appropriated and subsumed (and we have seen many cases of this in earlier chapters: hardware modifications, software modifications, ways of organizing groups, certain practices, hybridization of devices—all of these have been quickly "productized").

In the last decade, the problem of "what it means to live in the network society" has been widely investigated. In the critical literature, there is a broad consensus that digital information networks produce a certain monotonous reduction of networks to systems of control. Sober sociological accounts conclude: "It is arguable that the penetration of network technologies into workplaces, sites of production, the infrastructure of global commerce and the practice of warfare—in a phrase, their use as *technologies of systems control*—will ultimately be more important to the shape of the network society that their use as technologies of interpersonal and mass communication" (Barney 2004, 60).

This argument readily fits wireless networks. Yet any such conclusion is weakened to the extent that it still assumes that production (workplaces, sites of production) and consumption/communication ("interpersonal and mass communication") remain in principle separable. Similarly, critical media theory responses, such as Steven Shaviro's *Connected, or, What It Means to Live in the Network Society,* end by saying: "So this is what it means to live in the network society. We have moved out of time and into space. Anything you want is yours for the asking. . . . The one real innovation of the network society is this: now surplus extraction is at the center of consumption as well as production" (Shaviro 2003, 249).

The move out of time into space and the gratification of any desire through connectivity both pivot around the extraction of surplus value from "consumption and production." The movements initiate shifts in embodiment, subjectification, and value that crystallize distinctly in many of the practices, actions, maps, architectures, topologies, databases, and ideas of wirelessness.

Out of this, where would the radical empiricist researcher proceed? Much of this book has dwelt on the problem of what remains irreducible to the network and the network society. The air interfaces of wirelessness, I have argued, bring something irreducible to systems of control or networked productive consumers. Radical empiricism approaches problems differently from the standard sociological or ethnographic practice of "pursuing members' meanings" (Emerson, Fretz, and Shaw 1995). We can frame radical empiricism in relation to Rabinow's notion of assemblages, the matrices of experimentation out of which recent past and near futures emerge. Assemblages are themselves taken up in the elaboration of apparatuses that are more general. They feed into and are fed by problematizations—the large-scale events that organize differences and interplays of true and false. Yet radical empiricism also diverges from that problematizing process in some ways. Much of what I have had to say about wirelessness follows the "spiritual currents" Bergson (1934, 243) attributes to James: "We bathe, according to James, in an atmosphere traversed by large spiritual currents." "Spiritual currents" could be read in many ways. I have treated them as beliefs occasioning passional decisions about networks that generate actions, practices, ideas, verifications, and validations. Everything that billows out of the conjunctive envelope of wirelessness is provisional, often anticipatory, experimentally unstable, and subject to the liveness, momentousness, and force of belief.

In a sense everything I have been discussing here—the algorithms and antennae, the various overflows (spatial, thing, body, private-public), and the transitions between different versions of Wi-Fi or other wireless networks—concern how the passing of one experience into another, is subject to reorganization in wirelessness. A radical empiricist concept of experience touches on such questions at several points. In its insistent grounding of experience in transition, it goes to the heart of wirelessness. Transitions are conjunctive relations in the process of varying degrees of intimacy and proximity. Not only do the conjunctive relations of wirelessness as form of experience include a variety of overflows, those overflows are themselves the very sense and tendency of experience. Yet they themselves are not pure or aligned with each other. They are exposed to divergent forms of "verbalization." Wirelessness as a contemporary mode of experience is not pure in any sense. As a theoretical construct, it is not reducible to phenomenological, existential, or even psychological modes of understanding. As we have seen, it envelops "diverse processes," including those that are normally understood as belonging to objects, devices, gadgets, things, infrastructures, and thinking. If we look at some of the distinguishing features of those things, antennae and signal processing stand out as what make wireless media different from, say, gaming consoles or cameras (although these, of course, are becoming increasingly wireless). The wireless antennae and algorithms seek to generate certain conjunctive relations ("with," "to," "for") that hold experience together. They intensively reorder signals in the name of a connectivity that can tolerate interference or the presence of others. Yet, once we begin to explore wireless media in practice, it seems that the kinds of conjunctive relations they recruit are not easily controlled, corralled, or limited. They overflow in equivocal proximities—into other things, into living bodies, and across legal, physical, and social boundaries. These overflows all affect transitions. They constitute changes in the ways that transitions happen.

There is no ground for wirelessness, not even the mostly unfelt reaches of the electromagnetic spectrum. Instead, the radical empiricist account of experience allows us to say that the most intimate and most impersonal can sometimes come close to each other. Things and sensations are not at opposite poles of experience. This almost brings us full circle. We have glimpsed the mediatized, materialized, contested, commodified, politicized, normalized, and ignored kaleidoscopic cascade of changes associated with wireless networks. Many of these changes seek to connect or align what was previously separated or misaligned. But in almost every attempt to converge, they disturb the rankings of conjunctive relations between impersonal and personal, between remote and intimate. The flow of experience has to be reconfigured.

Notes

Chapter 1

1. The "wireless sniffing and monitoring" software *Kismet* runs on a laptop (Kershaw 2008). For instance, a Kismet scan for wireless devices in a train car typically shows several active Wi-Fi devices.

2. In many respects, the figure and practice of the network distinguish wirelessness today from the telegraphic wirelessness of the early twentieth century. The sensations of wireless movement and relation that excited so much interest a hundred years ago (for instance, in futurist writings on radio analyzed by Timothy Campbell (2006)) recapitulate themselves today in the much more dispersed or extensively distributed network forms.

3. Literary, sociological, anthropological, historical, and other critical studies concerning contemporary science and technology pursue a common question: How does technology change what it is to be human? (For the term *human*, other words such as *social*, *cultural*, *political*, *economic*, or simply *who we are* can be used.) Here, I want to resist any such framing. Among the many problems facing that form of questioning is this: How do you counteract what we might term "overrelevance"? Many high-profile technological changes such as mobile phones, a new drug for breast cancer, or a stem-cell treatment for multiple sclerosis are all too portentously relevant. In the case of the mobile phones, for instance, this abundant relevance makes it hard to think past the obvious impacts on patterns of mobility, modes of habitation, social practices of media use, and urban everyday life. Work on these social impacts, especially in relation to mobile phones, is not scarce, it is constantly growing and will likely continue as the various wireless technologies (3G, Wi-Fi, Bluetooth, WiMax) intermingle and hybridize. The social impacts of mobile phones are too great to ignore. In many respects, the transit time between technology and social impact on human life is too fast. To slow down enough to enter into the processes of change, other ways of constructing an engagement with change need to be found. One reason why this book does not have much to say about mobile phones or "smartphones" such as Apple's iPhone is that they are so glossily obvious

and relevant. However, it does take quite a strong interest in the increasing overlap between techniques used to construct and manage mobile phone infrastructures and Internet networks.

4. At this point, many readers might be feeling uneasy. Surely concepts of experience are just too philosophically depleted for further use. Aren't they so engrained with the figure of the human that any actual encounter with change glances off them? Yes and no. If wirelessness is a form of experience, it is not one that leaves the subject of experience, the one who experiences, unchanged. It does not leave experience itself unchanged, and because of that, the one who experiences, also changes. (The subject, or subjectivity, is always one who experiences something, even if only themselves as a subject.) It is on this point that James's radical empiricism can make a crucial difference. One of the basic questions of this book is how we make sense of changes in the fabric or texture of experience, without understanding that change in terms of pregiven notions of experience and its subjects. I am very much in sympathy with those strands of critical work on media and technology that call for accounts of human subjects and technological objects co-constituting hand in hand (see Mackenzie 2002). However, one of the interests in promoting an account based on experience and experiments in rapid change is to see how that encounter can generate specific changes, or channel inmixing of both living and nonliving things in forms of unexpected otherness or alterity.

5. The problem of overconnectedness has already been a topic of network theory. From the perspective of connectivity and stability, this phenomenon had already by been pursued by network sociologies and social network theorists (Van Dijk 2006, 187–188). Such discussions often rely on the formal studies and modeling of network connectivity found in the work of Stuart Kauffman (1995) or Albert-László Barabási and Duncan Watts (Barabási 2002; Newman, Barabási, and Watts 2006). They often conclude that too many connections in a network leads to lack of adaptive capacity or reduced ability to change. One could argue that certain wireless projects, such as the municipal wireless network in Philadelphia or San Francisco, foundered on the problem of too many connections. However, this structural analysis of overconnection as an intrinsic property of the network form largely renders illegible the practices of networking. It glosses over the work of making or removing connections, and slides over the differences in intensity of connection. As we will see, the making and breaking of network connections is a highly invested process, and kindles potent competition between several different technical standards, architectures, topologies, configurations, and perhaps more importantly, cultures of wireless networking. Network connections multiply as wirelessness intensifies. Complications at various levels—at the level of the technical standards and protocols that define wireless networks, at the level of attempts to control, configure, and map wireless networks, and at the level of attempts to expand or even generalize wireless networks globally (as in wireless development projects)—produce kinks and torsions in wirelessness.

6. This phrase comes from one of the reviewers of the manuscript of this book.

7. "Tout y est pris sur un même plan: idées, propositions, impressions, choses, individus, sociétés. L'expérience, c'est cet ensemble diffus, enchevêtré, de choses, de mouvements, de devenirs, de relations, sans distinction première, sans principe fondateur" (Debaise 2005, 104).

8. The point of this variation on James's account of the rationale of radical empiricism lies along similar lines to that proposed by Elizabeth Grosz (2005, 143) when she suggests that Bergson helps us reapprehend the thing: "to orient technology not so much to knowing and mediating as to experience and the rich interdeterminacy of duration, to a making without definitive end or goal."

While I turn to Bergson at several points in the following chapters, and specifically in the conclusion, James's work also emphasizes a superabundance of the real and experience. Bergson's (1934, 240) highly affirmative regard for James centers on just this point: "Whereas our motto to us is *only what is needed,* that of nature is *More than is needed,* too much of this, too much of that, too much of everything. Reality, as James sees it, is redundant and superabundant."

However, if Bergson's work has facilitated diverse and wide-ranging interventions in debates around audiovisual media such as cinema, video, and interactive or digital art (Hansen 2004; Rodowick 2001), James's work, much less known in media and cultural studies, stresses a specific set of relations that I find conducive to the problem of wireless technologies. James's emphasis on *conjunctive* relations lends much greater weight to the wirelessness of wireless technologies. Even in simply pointing to conjunctive relations, and bringing them into the flow of experience, radical empiricism moves in the direction set out by Grosz.

9. The word *consciously* evokes *unconsciously,* and can thereby switch on the paradigm of subjectivity. The complementary term best substituted here would be *nonconscious.*

10. For a fully developed account of James's political philosophy, see Ferguson 2007.

11. How many different aspects of wirelessness will we need to examine? The treatment in the following chapters marks out space, action, belonging, awareness, and globality as key facets of wirelessness. In choosing different scales of analysis for this book, I have attempted to remain alert to the problem discussed by historian of information technology Paul Edwards (2003) in an essay on infrastructures and modernity. Because we live "within multiple, linked infrastructures," we also "inhabit and traverse multiple scales of force, time and social organization" (p. 222). Confining analysis to any one scale (typically the micro for media, cultural, and social studies of uses and practices; the meso for analyses of institutions, organizations, and systems; the macro for studies of capitalism, modernity, etc.) jeopardizes the possibility of following these traversals, and jettisons any chance of finding out what transformations and translations come of it. Edwards argues that multiscale (micro-, meso-, and macroscale) analyses of technology, society, and nature are

needed. In contrast to the microemphasis found in social constructivist accounts, and the sometimes weighty static structures of macroscale accounts, radical empiricism has no pregiven commitment to any particular scale of analysis. It is constitutionally open to multiscale analysis by virtue of its highly composite treatment of experience.

12. For an extended discussion of Nancy on capital and Being, see Ross 2007.

13. Nancy's work derives from a French deconstructionist reading of Heidegger. The match between his style of writing and James's is, to say the least, rather poor. However, while Nancy's elliptical style and heavy hyphenation pose certain reading challenges, his intensive development of the conjunction "with" in framing of Being-with offers, it seems to me, a highly useful expression of the most extreme forms of capture of conjunctive relations. This will be particularly useful in understanding "global wireless" or "wireless worlds."

Another way to go here would be to examine the contemporary appropriation of intersubjective relations in communicative processes. This has been analyzed extensively in Marxist and post-Marxist political economy, cultural, and media theory, particularly in the work of Paolo Virno and Maurizio Lazzaratto (Lazzaratto 1998; Terranova 2000; Virno 2004).

14. Given all possible permutations, the following seven chapters could be read in a thousand or so different orders. The order in which I have arranged the chapters moves from the city to chipset/algorithm, to device, to networks, to media and markets, and then to wirelessness writ large in global development projects. But the chapters could be read in a different order: from small to large, from chip (chapter 3) to globe (chapter 7), via devices (chapter 3), network actions (chapter 4), media markets (chapter 5), and (chapter 2) cities. That order would reflect a gradually broadening theoretical argument that begins with minute sensations and moves to global awareness of others.

Chapter 2

1. A wireless-city literature is gradually beginning to take shape. See Forlano 2008 as well as Hampton and Gupti 2008.

2. The urban grid reappears, as the next chapter discusses, in the architectural layout of digital signal-processing chips, as well as in the algorithms used to shape wireless signals.

3. In *The Meaning of Truth,* James (1909, 45) writes: "To call my present idea of my dog, for example, cognitive of the real dog means that, as the actual tissue of experience is constituted, the idea is capable of leading into a chain of other experiences on my part that go from next to next and terminate at last in vivid sense-perceptions of a jumping, barking, hairy body."

The idea of a dog and real dogs appear often in James's writings about truth and knowledge. As a methodological point, in my handling of radical empiricism, I try to treat epistemological problems as problems of ambulation and tethers. Hence a dog off the leash in a crowded city street posed problems of coordination and direction, not of knowing that it is a dog. Given that the idea of wireless city responds to people moving in communication, the jumping, barking, hairy body in Leicester Square and wireless London are part of the same chain of experience.

4. WIFLY目前除了臺北市七大商業區、捷運全線、7-ELEVEN、星巴克咖啡、伊是咖啡等連鎖咖啡店，也包含全省各連鎖商店室內上網點，全台擁有近5,000個上網熱點，讓你擺脫有線束縛，隨時隨地無線暢遊，滿足職場與日常生活所需。

5. Belatedly, Beijing came to have a public wireless network. At the 2008 Beijing Olympics, Beijing launched itself as a wireless city, and only in a very limited version around important tourist and business precincts (see map on CECT—China Communications Ltd. 2008).

6. In her account of the contemporary network cultures, Tiziana Terranova (2004, 153–154) proposes:

In a network culture, the differentiating power of image flows achieves a kind of hydrodynamic status characterized by a local sensitivity to global conditions. . . . The result seems to be a political field that cannot be made to unite under a single signifier . . . or even under a stable consensus; while at the same time it cannot really be split off into separate segments with completely socio-cultural identities (even hybrid ones)—a space that is *common* without being *homogeneous* or even *equal*.

Terranova theorizes the "differentiating image flows" in the context of mass communication, and a mass audience's capacity to defy reason, analysis, or classification in the process of their responses. The "hydrodynamic status" of these image flows, while not discussed in detail, might be something like a turbulent flow, a flow of images in which many vortices, eddies, and cascades are seen. Something similar, I would argue, applies to wirelessness.

Chapter 3

1. But the signal-processing techniques encountered in Wi-Fi irradiate a spectrum of different technologies and sciences such as mobile phones, audiovisual media, bioinformatics and biometrics, and contemporary digital media.

2. Broadband wired networks such as ADSL (Asymmetric Digital Subscriber Loop) and cable modems also use the same digital signal-processing techniques.

3. "The scope of the event is part of its effects, of the problem posed in the future it creates. Its measure is the object of multiple interpretations, but it can be measured by the very multiplicity of these interpretations" (Stengers 2000, 67).

4. See Mackenzie 2009, 2008, for discussion of video signal-processing algorithms.

5. In Michel Serres's terms, they participate in the processing of collectives or world making, not by adding I's to each other, but by allowing partial withdrawal of an I (Serres and Schehr 1982, 228). This withdrawal of an I or subjectivity leaves behind heavy empirical and analytical obstacles.

6. In science and technology studies, researchers study science or technologies in the making in order to see how different interests or relations shape the construction of the algorithms included on the chip (FFT, R-S, Viterbi—some of which I discuss below). In fact, they usually leave the algorithms, equations, or functions aside in favor of "social dimensions" (Latour 1999, 111). As Isabelle Stengers (2005, 155) has argued, social constructivist analyses of technology in the making allow themselves a certain luxury of "deconstructing others' dreams."

7. In his account of Deleuze and Guattari's relation to science, Manuel Delanda (2002, 116) understands the discovery process in terms of literalization:

In epistemological terms to extract an ideal event from an actually occurring one is, basically, to define what is *problematic* about it, to grasp what about the event *objectively stands in need of explanation*. This involves discerning in the actual event what is relevant and irrelevant for its explanation, what is important and what is not. That is, it involves correctly grasping the *objective distribution of the singular and the ordinary* defining a well-posed problem.

While I find Delanda's reading of Deleuze and Guattari on science overly vigorous in some respects, it can be rendered pragmatically. The following discussion of signal processing could be read as extraction of "an ideal event." Isabelle Stengers (2005) puts the same point a different way. As she writes, "A philosophical 'counter-effectuation' would not be a strange 'Mime,' but would create by its own means what busy scientists so easily forget, namely the 'dignity of event' that makes them busy" (p. 158).

8. Humanities and social science work on the fast Fourier transform is hard to find, even though the FFT-IFFT couplet is the common mathematical basis of much contemporary digital image, video, and sound compression, and hence conditions most digital multimedia (in JPEG, MPEG, and MP3 file formats and in DVDs, for instance). In the early 1990s, Friedrich Kittler wrote an article that discussed it (Kittler 1993). His key point was largely that there is no real time in digital signal processing. The FFT works by defining a sliding window of time for a signal. It treats a complicated signal as a set of blocks that it lifts out of the time domain and transforms into the frequency domain. The FFT effectively plots an event in time as a graph in space. The experience of real time is epiphenomenal. In terms of the FFT, a signal is always partly in the future or the past. Although Kittler was not referring to the use of FFT in wireless networks, the same point applies—there is no real-time communication. While this point about the impossibility of real-time calculation was important to make during the 1990s, it seems well established now. I think an analysis of how FFT is used in wireless networks (but also in many wideband communication systems such as ADSL, DVB, and DAB) can do more than critique naive notions of real time.

9. Deleuze and Guattari (1994, 188) write of the scientific function, "It relinquishes the infinite, infinite speed, in order to gain *a reference able to actualize the virtual.*" In their terms, the FFT here is an attempt to actualize the virtuality of networks in urban or severe signal environments.

10. It is now also commonly used in information theory, speech recognition, keyword spotting, computational linguistics, and bioinformatics.

11. Probabilistic processes have been of great interest ever since population became a thinkable entity in the nineteenth century, something that could be counted and controlled (Thrift 2004a, 588). Probability and randomness have a number of different conceptual layers in the Viterbi algorithm. First, the algorithm regards all signals, sequences, or systems in probabilistic terms. In particular, it treats them as a hidden Markov model. A Markov model is a way of understanding stochastic processes. It models situations by assuming that the next state or event in the system is dependent only on the present state, not past states, and by assuming that the most probable next state can be known. For instance, the game of snakes and ladders can be modeled as a Markov process because the next move depends entirely on the present state of the board and on a roll of the dice. (More formally, the conditional probability distribution of next moves in snakes and ladders depends on a current state and roll of the dice.) Similarly, in convolutional coding, the next bit in a data stream relates to the preceding bits. The explanatory analogy supplied by *Wikipedia* under "Viterbi algorithm " is useful:

Assume you have a friend who lives far away and who you call daily to talk about what each of you did that day. Your friend has only three things he's interested in: walking in the park, shopping, and cleaning his apartment. The choice of what to do is determined exclusively by the weather on a given day. You have no definite information about the weather where your friend lives, but you know general trends. Based on what he tells you he did each day, you try to guess what the weather must have been like.

You believe that the weather operates as a discrete Markov chain. There are two states, "Rainy" and "Sunny," but you cannot observe them directly, that is, they are *hidden* from you. On each day, there is a certain chance that your friend will perform one of the following activities, depending on the weather: "walk," "shop," or "clean." Since your friend tells you about his activities, those are the *observations*. The entire system is that of a hidden Markov model (HMM).

You know the general weather trends in the area and you know what your friend likes to do on average. . . .

You talk to your friend three days in a row and discover that on the first day he went for a walk, on the second day he went shopping, and on the third day he cleaned his apartment. You have two questions: What is the overall probability of this sequence of observations? ("Viterbi Algorithm" 2008)

The explanation, symptomatically, invokes an everyday situation. It renders the core of the algorithm as a social situation. It renders the Viterbi algorithm thinkable as a set of questions about the weather.

12. Every algorithm, apart from so-called brute-force algorithms, contains a twist or kink that affects the flow of the computational process. (The fast Fourier transform is fast because it applies a "divide-and-conquer" strategy to synthesize or analyze the components of a signal.) The hidden Markov model is one twist or kink at the heart of the Viterbi algorithm. It treats the received signal as a set of states that correspond to a Markov model that cannot be observed directly. Via the hidden Markov model, the Viterbi algorithm turns the communication situation inside out. The combination of a known and finite set of system states and probability is turned inside out by the algorithm because it treats the Markov model as hidden. The object of making a hidden Markov model is to deduce the most probable sequence of internal states that could have given rise to the observed sequence of signals.

Chapter 4

1. I understand experiment here along the lines suggested by Isabelle Stengers (2000, 89): "This is the very meaning of the event that constitutes the experimental invention: *the invention of the power to confer on things the power of conferring on the experimenter the power to speak in their name.*"

This two-stage conferral of the power to speak in the name of something seems to me to succinctly express some of the essential tensions between different kinds of work done by wireless devices. Some of the techniques described in this chapter confer powers of speaking in the name of networks in the world; others resist conferring any such power.

2. Other symptoms would also include circuit bending for music and sound, dorkbot events (dorkbot 2008), hacklabs (London Hacklabs Collective 2008), and, indeed, the many art, activist, and community-based wireless network projects of 2001–2005.

3. "Progresser dan la connaissance . . . signifie passer d'un mauvais à un bon reseau" (Latour 2007, 28). (All translations from French are my own.) This statement appears in a discussion of James's radical empiricism in the context of science.

4. This can be traced back to Heidegger's (1967) distinction in *Being and Time* between two different modes of being: present-to-hand and ready-to-hand.

5. On this point, see Chun 2006 and Galloway 2004.

6. Actually, despite their proliferation, the hotspots have not, it seems, been very hot. The bar employees often do not know of the hotspot's existence. Many hotspots are rarely used due to their excessive cost and because they remain, ironically, relatively invisible and difficult to access (Frankston 2003).

7. The many attempts to federate or associate wireless devices with each other in networks or "wireless mesh" would be one main consequence. This will be discussed in the following chapter.

Chapter 5

1. The basic opposition between globalizing network and resistant locality seems to structure network society literature, as many commentators have remarked (Couldry 2004; Miller and Slater 2000).

2. "Chaque expérience est une action et chaque action a un center qui la désigne comme réalité singulère" (Debaise 2007, 13).

3. "On pourrait dire que l'invention concrétisante réalise un milieu techno-géographique . . . , qui est une condition de possibilité du fonctionnement de l'objet technique" (p. 55).

4. "L'adaptation-concrétisation est un processus qui conditionnne la naissance d'un milieu au lieu d'être conditionné par un milieu déjà donné; il est conditionné par un milieu qui n'existe que virtuellement avant l'invention; il y a invention parce qu'il y a un saut qui s'effectue et se justifie par la relation qu'il institue à l'intérieur du milieu qu'il crée" (p. 55).

5. Henri Bergson (1988, 171) understands "localization" of memory along usefully similar lines: "The work of localization consists, in reality, in a growing effort of expansion, by which memory, always present in its entirety to itself, spreads out its recollections over an ever wider surface and so ends by distinguishing, in what was till then a confused mass, the remembrances which could not find its proper place."

6. Some locative artworks such as *Loca* (Hemment et al. 2006), a work installed in city streets and art galleries to log and message passing Bluetooth-enabled mobile devices, addressed issues like surveillance and control. Such works I would argue sometimes shy away from the more challenging aspects of Tarkka's question about potentialities for thinking and acting.

7. Moreover, the enmeshing of the community wireless node database with providing services for consumers goes deeper. The NodeDB.com developers affirm their relation to commercial node databases by quoting someone else's response:

Is the source code available? (following was written by the FindU.com website author, for FindU.com website, a HAM mapping site. However this is the exact feeling of those that have contributed to the project, and couldn't have written it better if we tried.)

No. Commercialization is the primary reason I've not open-sourced my code. I've gotten a surprising number of requests from persons inside the GIS industry to either give or sell them my code, or consult for them. (NodeDB.com 2007)

This is a slightly confusing comment in some ways. The developers explain their decision not to release the source code using someone else's explanation for why they have decided not to release their source code. The justification offered by the author of the FindU.com website becomes the expression of a collective decision. What I find interesting in the comment is the acknowledgment that the node database has changed the way the developers think of their own actions: others may ask them for their work, or ask them to act as consultants on presumably commercial projects.

8. As Felix Guattari (1984, 135) writes, "One can always replace any pronoun with 'it,' which covers all pronominality, be it personal, demonstrative, possessive, interrogative or indefinite, whether it refers to adjectives or verbs."

In a sense, this chapter traces the process of that replacement in node databases.

9. For instance, as Wendy Chun (2006, 37) has argued, the Internet acquired a sociopolitical aura of the "medium of freedom" only by becoming cyberspace in the mid-1990s.

10. See the account of code as "abductive" in Mackenzie 2006.

11. As Marcel Mauss or Pierre Bourdieu do. Bourdieu (2000, 148) writes: "Action is neither 'purely reactive,' as in Weber's phrase, nor purely conscious and calculated. Through the cognitive and motivating structures that it brings into play (which always depend, in part, on the field, acting as a field of forces, of which it is the product), habitus plays its part in determining things to be done, or not to be done, the urgencies, etc., which trigger action."

12. Nancy (2000, 64) writes in relation to capital and "with": "The intuition buried in Marx's work is undoubtedly located in the following ambivalence: at one and the same time, capital exposes the general alienation of the proper—which is the generalized disappropriation, or the appropriation of misery in every sense of the word—*and* it exposes the stripping bare of the *with* as a mark of Being, or as a mark of meaning."

My use of Nancy here attempts to identify "with" as a conjunctive relation. If being-with-another is the only irreducible "mark of meaning," the one thing that cannot be alienated, then careful analysis of forms of "with" might be valuable.

13. This point also applies to things. As Simondon (1989, 52) writes, "The technical object is at the meeting point of two milieus, and it must be integrated into two milieus at once" (L'objet technique est au point de rencontre de deux milieux, et il doit être intégré aux deux milieux à fois).

Chapter 6

1. Sometimes the listing of places goes in strange directions. Wi-Fi has been the object of some slightly absurd visioning or "hype." The somewhat confusingly named Wi-Fi TV is a "new generation social software Internet TV" (Wi-iFi TV 2008). In other words, it is a website where registered users can subscribe to many (hundreds, it is claimed) "TV stations." Why name a TV-oriented website "Wi-Fi TV"? Somehow the name was meant to convey an idea of convergence between broadcast and network media. Should we dismiss this as just another of the ceaseless attempts

to sell services on the Internet by harnessing desires and beliefs, albeit this time a bit misleadingly?

2. The network society of the 1980s and 1990s involved moving ever-mounting quantities of information between points on the relatively fixed networks. The development of wireless networks, however, changes the stakes, according to Mitchell (2003). It is a question of "reactivating" public spaces by bending network fixtures to suit different kinds of local movements:

> In general, as these transformations of public space illustrate, there is a strong relationship between prevailing network structure and the distribution of activities over public and private places. . . . And where networks go wireless, they mobilize activities that had been tied to fixed locations and open up ways of reactivating urban public space; the home entertainment center reemerges as the Walkman, the home telephone as the cellphone, and the computer as the laptop. (p. 158)

While these changes are often presented as ubiquitous, it is immediately obvious that they are unevenly distributed, not just spatially, but in relation to other "activations" of social spaces—for instance, as sites of consumption.

3. Several years later, it seemed that Intel's hype had paid off: "Intel has thanked strong demand for its Centrino [Wi-Fi-equipped] laptop processors for a 29% rise in quarterly profits. The world's largest chip maker saw net income for its first quarter ending 2 April increase to $2.2bn (£1.1bn), against $1.7bn a year earlier" (BBC 2005).

4. See chapter 7 for further discussion of states and wirelessness.

5. "C'est dans les métamorphoses et les variations de l'action de subjectivation et non les métamorphoses et les variations de valuer qu'il faut chercher la dynamique immanente du capitalisme."

6. This is consistent too, at least in key respects, with Jean-Luc Nancy's account of how the appropriating processes of capital inadvertently expose relations of coappearing or "with." In this context, the list literally puts variations in subjectification "with"each other, despite the fact that they might not belong together.

7. The recently reawakened interest in Tarde surveyed in Barry and Thrift 2007 emphasizes this point: the basic "social fact" is fluxes in relation.

8. "C'est par l'affectionabilité, par le sentir pur, fonds commun de désirs et de croyances, que Tarde décrit la différence à la fois de nature et de degré entre les activités de creation et de reproduction" (Lazzarato 2002, 21).

9. "Le pensée et la chose, le sujet et l'objet ne sont pas des entités séparées, ou des substances. Ils sont des *modes* irréductiblement temporels *de relation de l'expérience à elle-même.*"

10. The general point about how technologies become background or infrastructural is well known from social studies of technology (Bowker and Star 1999), but the process of making something visible in some forms so that it can become

invisible in others is much less well understood. Here, the process of sorting experience as inner and outer, and equivocations in this sorting, play a key role.

Chapter 7

1. Many of the most visible wireless development projects are actually public-private partnerships. These partnerships take many forms. Sometimes these partnerships themselves are highly complex networks. For instance, NetHope describes itself as

[a] new generation Information and Communications Technology consortium of 20 of the international community's leading non-governmental organizations. NetHope is dedicated to the best use of available technology resources by its member agencies in order to improve their ability to deliver community benefit in the developing world. It does this by working as a highly collaborative team, sharing ICT knowledge, solving common problems, creating industry relationships to support the public benefit work of all of its members, and educating members as well as the wider community of NGOs. (NetHope LLC 2007)

While it is hard to judge what "new generation" means in the context of development, NetHope seeks to federate NGOs into a consortium that shares ICT kits for use in relief efforts for disasters and war. They offer advice to NGOs on how to establish network connectivity among teams, and communication between the disaster or war zone and other parts of the planet. While they do use wireless technologies such as Wi-Fi, aid agencies mainly use satellite connections equipment to communicate via the four VSAT (Very Small Aperture Terminal) satellites owned by Inmarsat. As Télécoms Sans Frontières (2007) describes their own equipment, "The Inmarsat BGan [Broadband Global Area Network] is the latest innovation in satellite communication equipment. It permits 10 computers to the Broadband Internet and also offers voice and fax services. The BGan is light (less than 4 kg) and mobile (deployable within minutes)."

The satellite networks invoked here were first developed for shipping and mineral exploration use, but are now reconfigured as part of a "broadband global area network." Each term of this Bgan contraction makes a claim to do something quite surprising since it brings together broadband (a claim to high-volume communication), global (as in covering the earth), area (now expanded to include the globe), and network (the standard model-figure-practice of communication).

2. "Nous ne dirons jamais que le capitalisme est <<pragmatique>>. C'est plutôt l'antipragmatique par excellence, car l'entreprise de redéfinition systématique qui lui est associée n'est pas obligée par une quelconque vérification, ni par une pensée soucieuse de consequences" (Pignarre and Stengers 2005, 30).

3. Jean-Luc Nancy (2000, 69) writes: "The being-in-itself of 'society' is the network and cross-referencing [le renvoi mutuel] of co-existence, that is, of coexistences. That is why every society gives itself its spectacle and gives itself as spectacle, in one form or another."

4. See the chapters on "one or many worlds" in *Some Problems of Philosophy* (James 1911). For a more general framing of James in the context of political theory, see Ferguson 2007.

5. See Fillip and Foote 2007 for an extensive analysis.

6. See discussion of Unified Modeling Languages in Mackenzie 2006.

7. Télécoms béninois: anarchie totale, mesures draconiennes annoncées

Une étude commanditée par l'Etat du Bénin dans le secteur des télécommunications dans le pays a fait "l'amer constat" de la grande anarchie qui y prévaut, incitant le gouvernement à annoncer une série de mesures drastiques, a appris Ouestafnews de source officielle.

L'étude révèle qu'un total de "47 opérateurs sur 50 visités," exercent "tout ou partie de leurs activités en violation des textes" privant ainsi l'Etat béninois "d'importantes recettes" et provoquant "une forte hémorragie financière au préjudice de l'Etat béninois," affirme un communiqué du Conseil des ministres parvenu à Ouestafnews. . . .

L'Etat, selon le communiqué va également suspendre tous les "arrêtés portant autorisation de prestation de services de télécommunications telles que la Voix sur IP, la boucle locale radio, le wifi, le wimax, l'adsl, les cartes prépayées à l'exception des fournisseurs d'accès internet et des opérateurs de cybercafé qui exploitent légalement leurs activités."

Les conditions de la récente "mutation des Telecel vers Moov" (opérateurs privés) seront également examinées, promet le gouvernement béninois qui annonce plusieurs autres mesures dont notamment le démantèlement "sans délai" d'installations techniques opérées sans autorisation, le relèvement des prix de licences accordées de "manière fantaisiste." (Ouestaf 2007)

8. "Wireless devices shall not be used for storing, processing, or transmitting classified information without explicit written approval of the cognizant DAA [Designated Approving Authority]. If approved by the DAA, only assured channels employing National Security Agency (NSA)-approved encryption shall be used to transmit classified information. Classified data stored on PEDs must be encrypted using NSA-approved encryption consistent with storage and treatment of classified information" (Wolfowitz 2004, 3).

Chapter 8

1. "Nous baignons, d'aprés James, dans une atmosphére que traversent de grands courants spirituels" (Bergson 1934, 243).

2. This is because most mobile devices unpredictably transit in and out of dormancy in order to extend battery life. Each time they rejoin the network, a series of signals have to be exchanged to set up a connection.

3. "La reàlité coule; nous coulons avec elle; et nous appelons vraie toute affirmation qui, en nous dirigeant à travers la reàlité mouvante, nous donne prise sur elle et nous place dans de meilleures conditions pour agir."

4. This convergence runs in both directions between James and Bergson. For instance, in *The Pluralistic Universe*, James (1909, 231–236) discusses at length how

for Bergson's ideas and concepts themselves are of practical value. See Ferguson 2007, chapter 4, for an account of the bilateral exchanges between Bergson and James.

5. Much work in feminism, cultural theory, media studies, and film studies filters Bergson via Gilles Deleuze (Boundas 1996; Grosz 2005; Mullarkey 1999; Rodowick 2001).

6. My discussion draws heavily on Deleuze's (1988) treatment of the method of intuition in *Bergsonism*, and runs it through James's radical empiricism. How does intuition work as a precise philosophical method rather than as "feeling, inspiration, [or] disorderly sympathy?" (Deleuze 1988, 13). Precision is a live issue for certain aspects of radical empiricism too, so the steps of Deleuze's analysis, and his extraction of "rules" of method from Bergson, are of interest. A full analysis of the method of intuition in contrast to radical empiricism demands a long philosophical note. Here I take up specific results that Deleuze derives in the course of his discussion of precision.

7. Bergson (1934, 23) writes: "Mais si l'on commence par écarter les concepts déjà faits, si l'on se donne une vision direct de réel, si l'on subdivise alors cette réalité en tenant compte de ses articulations, les concepts nouveaux qu'on devra bien former pour s'exprimer seront cette taillés à l'exacte mesure de l'objet."

References

Abbate, Janet. 2000. *Inventing the Internet*. Cambridge, MA: MIT Press.

ABC. 2005. *Access to Broadband Campaign*. http://www.abcampaign.org/.

ABIResearch. 2007. *More than a Billion Wi-Fi Chipsets to Ship in 2012*. February 13. http://www.abiresearch.com/abiprdisplay.jsp?pressid=809.

Adly, Hassan. 2004. *1KM, 2Mbps, 802.11b Wireless Link Using Linksys WAP11 + Yagi, in Hurghada, Egypt*. http://www.d128.com/wireless/.

Airjaldi. 2008. *Empowering Communities through Wireless Networks*. August 2. http://drupal.airjaldi.com/.

Aitken, Hugh G. J. 1985. *The Continuous Wave: Technology and American Radio, 1900–1932*. Princeton, NJ: Princeton University Press.

Akma. 2005. *So Weirdly Wrong*. http://akma.disseminary.org/archives/001518.html.

Albert, Saul. 2003. The Copenhagen Interpolation. *Mute Culture and Politics After the Net* 26:6–7.

Corp, Alcatel-Lucent. 2008 *Optimize Wireless Networks for Data*. http://www.alcatel-lucent.com/wps/portal/NewsFeatures/Detail?LMSG_CABINET=Docs_and_Resource_Ctr&LMSG_CONTENT_FILE=News_Features/News_Feature_Detail_000308.xml.

Andrejevic, Marc. 2007. Ubiquitous computing and the digital enclosure movement. *Media International Australia* 125:106–117.

Appadurai, Arjun. 1996. *Modernity at Large: Cultural Dimensions of Globalization*. Minneapolis: University of Minnesota Press.

Asadoorian, Paul, and Larry Pesce. 2007. *Linksys WRT54G: Ultimate Hacking*. Burlington, MA: Syngress.

Barabási, Albert-László. 2002. *Linked: The New Science of Networks*. Cambridge, MA: Perseus.

Barken, Lee. 2004. *How Secure Is Your Wireless Network? Safeguarding Your Wi-Fi-LAN.* Upper Saddle River, NJ: Prentice Hall.

Barney, Darin David. 2004. *The Network Society: Key Concepts.* Cambridge, MA: Polity Press.

Barry, Andrew. 2001. *Political Machines: Governing a Technological Society.* London: Athlone.

Barry, Andrew, and Nigel Thrift. 2007. Gabriel Tarde: Imitation, invention and economy. *Economy and Society* 36 (4):509–525.

BassStation. 2004. *About BassStation.* http://www.bass-station.net/index.php?photos.

BBC. 2003. *UN Urges Wi-Fi for All.* September 12. http://news.bbc.co.uk/1/hi/technology/3025734.stm.

BBC. 2005. *Strong Sales Boost Intel Profits.* April 19. http://news.bbc.co.uk/1/hi/business/4461903.stm.

BBC. 2007. *Wi-Fi: A Warning Signal.* http://news.bbc.co.uk/1/hi/programmes/panorama/6674675.stm

BBC News—Technology. 2005. *Surf the Net While Surfing Waves.* http://news.bbc.co.uk/1/hi/technology/3812357.stm.

Belson, Ken. 2006. What if they built an urban wireless network and hardly anyone used it? *New York Times,* June 26. http://www.nytimes.com/2006/06/26/technology/26taipei.html?pagewanted=2&ei=5088&en=8a1b4842ada0d065&ex=1308974400&partner=rssnyt&emc=rss.

Bergson, Henri. 1934. *La pensée et le mouvant; essais et conférences, Bibliothèque de philosophie contemporaine.* Paris: F. Alcan.

Bergson, Henri. 1988. *Matter and Memory.* New York: Zone Books.

Best, Jo. 2004. *BMW and HP to Make Wi-Fi Beemers for High-Flying Execs.* May 5. http://networks.silicon.com/wifi/0,39024669,39120476,00.htm.

Biever, Celeste. 2005. *Virtual Fences to Herd Wi-Fi Cattle.* http://www.newscientist.com/article.ns?id=dn5079.

Black, Paul E. 2006. *Manhattan Distance.* Dictionary of Algorithms and Data Structures.http://www.itl.nist.gov/div897/sqg/dads/HTML/manhattanDistance.html

Blank, Martin. 2009. Preface. *Pathophysiology* 16 (2–3):67–69.

Bleecker, Julian. 2006. Locative media: A brief bibliography and taxonomy of Gps-enabled locative media. *Leonardo Electronic Almanac* 14 (3):11.

Bleecker, Julian. 2008. *wifi.Bedouin*. http://www.techkwondo.com/projects/bedouin/WiFiBedouin.pdf.

Boccuzzi, Joseph. 2008. *Signal Processing for Wireless Communications*. New York: McGraw-Hill.

Boundas, Constantin. 1996. Deleuze-Bergson: An ontology of the virtual. In *Deleuze: A Critical Reader*, ed. P. Patton. Oxford: Blackwell.

Bourdieu, Pierre. 2000. *Pascalian Meditations*. Stanford, CA: Stanford University Press.

Boutin, Paul. 2004. *Feds Label Wi-Fi a Terrorist Tool*. http://www.wired.com/gadgets/wireless/news/2002/12/56742.

Bowker, Geoffrey C. 1994. Information mythology: The world of/as information. In *Information Acumen: The Understanding and Use of Knowledge in Modern Business*, ed. L. Bud-Frierman. London: Routledge.

Bowker, Geoffrey C., and Susan Leigh Star. 1999. *Sorting Things Out: Classification and Its Consequences*. Cambridge, MA: MIT Press.

Broadcom Corporation. 2007. *BCM4325 Low-Power 802.11a/b/g with Bluetooth® 2.1 + EDR and FM Product Brief* 2007. http://www.broadcom.com/products/Bluetooth/Bluetooth-RF-Silicon-and-Software-Solutions/BCM4325.

Broadcom Corporation. 2008. *Broadcom in Your Life*. http://www.broadcom.com/flash/broadcom_life/index.php?section=bs_section.

Bryant, Park. 2004. *The Bryant Park Wireless Network*. http://www.bryantpark.org/amenities/wireless.php.

Bureau d'Etudes. 2008. *Electro-Magnetic Propaganda: The Statement of Industrial Dogma*. 2007. http://rixc.lv/waves/en/txt06.html.

Callon, Michel, Cécile Méadel, and Vololona Rabeharisoa. 2002. The economy of qualities. *Economy and Society* 31 (2):194–217.

Callon, Michel, Andrew Barry, and Don Slater. 2002. Technology, politics and the market: An interview with Michel Callon. *Economy and Society* 31 (2):285–306.

Callon, Michel, Cecile Meadel, and Vololona Rabeharisoa. 2005. The economy of qualities. In *The Technological Economy*, ed. A. Barry and D. Slater. London: Routledge.

Campbell, Timothy C. 2006. *Wireless Writing in the Age of Marconi. Electronic Mediations*. vol. 16. Minneapolis: University of Minnesota Press.

Castells, Manuel. 1996. *The Rise of the Network Society*. Oxford: Blackwell.

Castells, Manuel. 2004. Space of flows, space of places: Materials for a theory of urbanism in the information age. In *Cybercities Reader*, ed. S. Graham. London: Routledge.

Castells, Manuel, Mireia Fernandez-Ardevol, Jack Linchuan-Qiu, and Araba Sey, eds. 2007. *Mobile Communication and Society: A Global Perspective.* Cambridge, MA: MIT Press.

Castoriadis, Cornelius. 1987. *The Imaginary Institution of Society.* Cambridge, MA: Polity Press.

CECT—China Communications Ltd. 2008. *Wireless City—Beijing* 2008. http://www .bj.wicity.cn/1/en/index.ph.

Chan, Sarah. 2007. *Model for the World: Digital Divide Closed in Cambodian Villages Where E-mail Is Delivered by Wi-Fi on a Motorbike.* http://slashdot.org/comments.pl? sid=94960&threshold=1&commentsort=0&tid=137&tid=193&mode=thread& cid=8143055

Chun, Wendy Hui-Kyong. 2006. *Control and Freedom: Power and Paranoia in the Age of Fiber Optics.* Cambridge, MA: MIT Press.

City of London Corporation. 2008. *City of London Goes Wireless: Launch of Europe's Most Advanced WiFi Network.* http://www.cityoflondon.gov.uk/Corporation/media _centre/files2007/73_07.htm.

Clement, Andrew, and Amelia Bryne Potter. 2008. Saving Toronto Hydro Telecom's One Zone project from itself: Alternative models for urban public wireless infrastructure. *Journal of Community Informatics* 4 (1). http://ci-journal.net/index.php/ciej/ article/view/472/421.

Coeur d'Alene Tribe. 2005. *Wireless ISP with Long Name but Big Ambitions.* http:// www.cdatribe-nsn.gov/tcbgp/.

Cohen, Randy. 2004. The ethicist: Wi-Fi fairness. *New York Times,* February 8. http:// www.nytimes.com/2004/02/08/magazine/08ETHICIST.html?ei=5062&en=2ee8eb4c df5d2976&ex=1076821200&pagewanted=print&position=

Cohn, David. 2005. *Pandas Don't Need Internet.* http://www.wired.com/news/ business/0,1367,65500,00.html?tw=wn_tophead_11.

Consume. 2003. *Consume the Net.* http://www.consume.net/index2.php.

Consume. 2008. *Node Info for Mildmay Grove.* http://nodedb.consume.net/ node_info.php?node_id=2011.

Convolutional coding. 2009. *Wikipedia.* http://en.wikipedia.org/wiki/ File:Convolutional_code_trellis_diagram.png.

Convolutional encoder diagram. 2007. *Wikipedia.* http://upload.wikimedia.org/ wikipedia/en/3/3c/Convolutional_encoder.png.

Corbridge, Stuart. 2007. The (im)possibility of development studies. *Economy and Society* 36 (2):179–211.

Cormen, Thomas H. 1990. *Introduction to Algorithms*. Cambridge, MA: MIT Press.

Couldry, Nick. 2004. The productive "consumer" and the "dispersed" citizen. *International Journal of Cultural Studies* 7 (1):21–32.

Cox, John. 2009. *Police in India Sweep for Unsecured Wi-Fi Networks*. Network World. http://www.networkworld.com/news/2009/011509-mumbai-police-unsecured-wifi-networks.html?fsrc=netflash-rss.

Cradlepoint. 2009. *PHS300 Personal WiFi Hotspot | Cradlepoint Technology*. http://www.cradlepoint.com/products/phs300-personal-wifi-hotspot.

Creamer Media. 2007. *Altech to Launch Broadband Services in Rwanda by Year-End*. June 27. http://www.engineeringnews.co.za/article.php?a_id=111716.

Debaise, Didier. 2005. Un pragmatisme des puissances. *Multitudes* 3 (22):103–110.

Debaise, Didier. 2007. *Vie et Expérimentation Peirce, James, Dewey*. Paris: Librairie Philosophique J. Vrin.

Delanda, Manuel. 2002. *Intensive Science and Virtual Philosophy*. London: Continuum.

Deleuze, Gilles. 1988. *Bergsonism*. Trans. H. Tomlinson. New York: Zone Books.

Deleuze, Gilles. 2001. *Difference and Repetition*. Trans. P. Patton. *Athlone Contemporary European Thinkers*. London: Continuum.

Deleuze, Gilles, and Félix Guattari. 1994. *What Is Philosophy? European Perspectives*. New York: Columbia University Press.

Derrida, Jacques. 1973. *Speech and Phenomena, and Other Essays on Husserl's Theory of Signs*. *Northwestern University Studies in Phenomenology & Existential Philosophy*. Evanston, IL: Northwestern University Press.

Doctorow, Cory. 2005. *Someone Comes to Town, Someone Leaves Town*. New York: Tor Books.

Dorkbot. 2008. *dorkbot people doing strange things with electricity* http://dorkbot.org/.

Dourish, Paul, and Genevieve Bell. 2007. The infrastructure of experience and the experience of infrastructure: Meaning and structure in everyday encounters with space. *Environment and Planning. D, Society & Space* 34:414–430.

Duffy, Carolyn Marsan. 2009. Disney finds "a whole new world" in WiFi-powered rides—802.11n, Disney, wi-fi. *PC World*. http://www.pcworld.idg.com.au/article/313735/disney_finds_whole_new_world_wifi-powered_rides.

Dunne, Anthony. 2005. *Hertzian Tales: Electronic Products, Aesthetic Experience, and Critical Design*. Cambridge, MA: MIT Press.

Edney, Jon, and William A. Arbaugh. 2004. *Real 802.11 Security: Wi-Fi Protected Access and 802.11i.* Boston: Addison-Wesley.

Edwards, Paul N. 2003. Infrastructure and modernity: Force, time, and social organization in the history of sociotechnical systems. In *Modernity and Technology*, ed. T. J. Misa, P. Brey, and A. Feenberg. Cambridge, MA: MIT Press.

Emerson, Robert M., Rachel I. Fretz, and Linda L. Shaw. 1995. *Writing Ethnographic Fieldnotes. Chicago Guides to Writing, Editing, and Publishing.* Chicago: University of Chicago Press.

e.ngo.org. 2008. *Spectral Ecology.* http://ellipsetours.free.fr/wikiblog/wakka.php?wiki=Blog-20070611225107-Projets.

Espiner, Tom. 2007. *Wi-Fi Hack Caused TJ Maxx Security Breach.* ZDNet UK. http://news.zdnet.co.uk/security/0,1000000189,39286991,00.htm

Federal Communications Commission. 2004. *O3L-PT-03-MN WIRELESS DESKTOP EDGE MOUSE.* https://fjallfoss.fcc.gov/oetcf/tcb/reports/Tcb731GrantForm.cfm?mode=COPY&RequestTimeout=500&tcb_code=&application_id=313181&fcc_id=O3L-PT-03-MN.

Ferguson, Kennan. 2007. *William James: Politics in the Pluriverse, Modernity and Political Thought.* Lanham, MD: Rowman & Littlefield.

Fillip, Barbara, and Dennis Foote. 2007. *Making the Connection: Scaling Telecenters for Development.* Washington, DC: Academy for Educational Development.

Solutions, First Mile. 2007. *Hybrid Real-Time, Store-and-Forward WiFi Mesh in Kigali Rwanda.* http://www.firstmilesolutions.com/projects.php.

Fisher, Melissa S., and Greg Downey. 2006. *Frontiers of Capital: Ethnographic Reflections on the New Economy.* Durham, NC: Duke University Press.

Flickenger, Rob. 2003. *Wireless Hacks 100 Industrial-Strength Tips & Tools.* Sebastapol, CA: O'Reilly & Associates Inc.

Flickenger, Rob, Corinna Elektra Aichele, Carlo Fonda, Jim Forster, Ian Howard, Tomas Krag, and Marco Zennaro. 2006. *Wireless Networking in the Developing World.* wndw.net eBook.

FMFI. 2008. *Wireless Africa Workshop.* July 1. http://www.fmfi.org.za/wiki/index.php/Wireless_Africa_Workshop.

FON. 2008. *Make Money With Your WiFi.* http://www.fon.com/en/info/makeMoney.

FON. 2006. *What Is Fon?* http://en.fon.com/info/what-is-fon.php.

Forlano, Laura. 2008. Anytime? Anywhere? Reframing debates around municipal wireless networking. *Journal of Community Informatics.* Special Issue: Wireless Networking for Communities, Citizens and the Public Interest 4 (1).

Frankston, Bob. 2004. *Hotspots, Cold Cells*. http://www.frankston.com/public/writing.asp?name=HotCold.

Fuller, Matthew. 2005. *Media Ecologies: Materialist Energies in Art and Technoculture*. Cambridge, MA: MIT Press.

Fuller, Matthew. 2008. *500-Slogans*. http://www.spc.org/fuller/fiction/500-slogans/.

Gabriel, Caroline. 2009a. *AT&T and Orange Highlight Mobile Data Boom—Rethink Wireless*. http://www.rethink-wireless.com/index.asp?article_id=1379.

Gabriel, Caroline. 2009b. *Back Office Vendors Gear Up to Support a Trillion Devices*. May 11. http://www.rethink-wireless.com/?article_id=1355&qtabs=99999.

Galloway, Alexander R. 2004. *Protocol: How Control Exists After Decentralization*. *Leonardo Series*. Cambridge, MA: MIT Press.

Galloway, Alexander R., and Eugene Thacker. 2007. *The Exploit: A Theory of Networks*. Minneapolis: University of Minnesota Press.

Galloway, Anne, and Mathew Ward. 2006. Locative media as socialising and spatialising practice: Learning from archaeology. *Leonardo Electronic Almanac* 14 (3–4):12.

Gardner, W. David. 2007. *A WiMax Network Goes Live in Tanzania*. WiMax InformationWeek. http://www.informationweek.com/news/mobility/showArticle.jhtml?articleID=203101044.

Ghadialy, Zahid. 2009. *All Data Packets Are Not Created Equal*. http://3g4g.blogspot.com/2009/07/all-data-packets-are-not-created-equal.html.

Gibson, Owen. 2002. Betting on Flip: Nicholas Negroponte is a digital pathfinder. *The Guardian*, July 15, 38.

Gitman, Yury. 2004. *Magic Bike: Wireless Access Bike*. http://magicbike.net/about.html.

Glasner, Joanna. 2004. *Trailer Parks Convert to Wi-Fi*. http://www.wired.com/news/business/0,1367,58784,00.html.

Gomes, Pete. 2004. *Welcome to Park Bench TV!* http://parkbenchtv.org/frameindex.html.

Google Inc. 2008. *What Is the Google Maps API?* http://code.google.com/apis/maps/index.html.

Graham, Stephen, and Simon Marvin. 2001. *Splintering Urbanism: Networked Infrastructures, Technological Mobilities and the Urban Condition*. London: Routledge.

Graham, Stephen, and Nigel Thrift. 2007. Out of order: Understanding repair and maintenance. *Theory, Culture & Society* 24 (3):1–25.

Greene, Kate. 2008. *Long-Distance Wi-Fi*. March 18. http://www.technologyreview
.com/Infotech/20432/?nlid=945&a&a=f.

Grosz, Elizabeth A. 2005. *Time Travels: Feminism, Nature, Power, Next Wave*. Durham,
NC: Duke University Press.

Groth, Duane. 2006. *Sydney Wireless*. http://www.sydneywireless.com/.

Grubesic, T. H., and A. T. Murray. 2004. "Where" matters: Location and Wi-Fi access.
Journal of Urban Technology 11 (1):1–28.

Guattari, Félix. 1984. *Molecular Revolution: Psychiatry and Politics. Harmondsworth*.
Middlesex, England: Penguin.

Hammersley, Ben. 2004. *Working the Web: Warchalking*. http://www.guardian.co.uk/
online/story/0,3605,748499,00.html.

Hampton, Keith N., and Neeti Gupti. 2008. Community and social interaction in
the wireless city: Wi-fi use in public and semi-public spaces. *New Media & Society* 10
(6):831–850.

Hannerz, Ulf. 1992. *Cultural Complexity: Studies in the Social Organization of Meaning*.
New York: Columbia University Press.

Hansen, Mark B. N. 2004. *New Philosophy for New Media*. Cambridge, MA: MIT Press.

Heidegger, Martin. 1967. *Being and Time*. Oxford: Blackwell.

Hellweg, Eric. 2005. *Wi-Fi for the Masses*. August 19. http://www.technologyreview
.com/InfoTech/wtr_14705,258,p1.html.

Hemment, Drew, John Evans, Theo Humphries, and Mika Raento. 2008. *Locative
Media and Pervasive Surveillance: The Loca Project*. http://www.drewhemment
.com/2006/locative_media_and_pervasive_surveillance_the_loca_projectby_drew
_hemment_john_evans_theo_humphries_mika_raento.html.

Herman, Wendy. 2007. *Wireless Mesh Good Fit for M-Taiwan Mobility Initiative*. http://
www.nortel.com/corporate/pressroom/feature_article/2005b/05_25_05_wireless
_mesh.html.

Hero, Anonymous. 2004. *Plant Yourself in the Lemon Tree*. http://www.warchalking.
org/section/pics.

Hian, Chua Hou. 2007. *AsiaMedia: Singapore: Wi-Fi Thief's Sentence Lauded as "Practical."* http://www.asiamedia.ucla.edu/article.asp?parentid=62202.

HiveNetworks. 2006. *About Hive Networks*. http://www.hivenetworks.net/mediawiki/
index.php/HiveNetworks:About.

Hong, Sungook. 2001. *Wireless: From Marconi's Black-Box to the Audion*. Cambridge,
MA: MIT Press.

Hume, Mick. 2006. *Wi-Fi Phobia: It Makes Me Sick.* November 24. http://www
.timesonline.co.uk/tol/comment/columnists/mick_hume/article647886.ece.

Husserl, Edmund. 1965. *Cartesian Meditations.* The Hague: M. Nijhoff.

Hutchins, Edwin. 1995. *Cognition in the Wild.* Cambridge, MA: MIT Press.

HyperlinkTech. 2007. *Innovative Solutions for Wireless Communications.* http://hyper-
linktech.com/.

ICTD2007. 2007. *IEEE/ACM International Conference on Information and Communication Technologies and Development.* http://research.microsoft.com/workshops/
ictd2007/.

IEEE. 1999a. *IEEE Std 802.11a-1999 Part 11: Wireless LAN Medium Access Control
(MAC) and Physical Layer (PHY) Specifications: High Speed Physical Layer in the 5 Ghz
Band.* New York: Institute of Electrical and Electronics Engineers, Inc.

IEEE. 1999b. *IEEE Std 802.11b-1999 Part 11: Wireless LAN Medium Access Control
(MAC) and Physical Layer (PHY) Specifications: Higher-Speed Physical Layer Extension in
the 2.4 GHz Band.* New York: Institute of Electrical and Electronics Engineers, Inc.

IEEE. 2003. *IEEE Std 802.11g-2003 Part 11: Wireless LAN Medium Access Control (MAC)
and Physical Layer (PHY) Specifications: Amendment 4: Further Higher Data Rate
Extension in the 2.4GHz Band.* New York: Institute of Electrical and Electronics
Engineers, Inc.

IEEE. 2004. *IEEE Standard for Local and Metropolitan Area Networks Part 16: Air Interface
for Fixed Broadband Wireless Access Systems.* New York: Institute of Electrical and
Electronics Engineers, Inc.

IEEE. 2007. *IEEE Standard for Information Technology— Telecommunications and Infor-
mation Exchange between Systems— Local and Metropolitan Area Networks— Specific
Requirements Part 11: Wireless LAN Medium Access Control (MAC) and Physical Layer
(PHY) Specifications.* New York: Institute of Electrical and Electronics Engineers, Inc.

Île Sans, Fil. 2008. *Sans Fil Wireless.* http://www.ilesansfil.org/tiki-index.php.

Instructables. 2009. *Old Dog, New Tricks: Mod a Linksys WRT54G-Series Router.* http://
www.instructables.com/id/Old-Dog,-New-Tricks:-Mod-a-Linksys-WRT54G-series-R/.

Intel Corporation. 2004a. *Discover the Freedom and Flexibility of Mobile Computing.*
http://www.intel.com/unwire/.

Intel Corporation. 2004b. *Intel Announces Centrino™ Mobile Technology Brand Name.*
http://www.intel.com/pressroom/archive/releases/20030108corp.htm.

International Telecommunication Union. 2009. *Free Statistics.* http://www.itu.int/
ITU-D/ict/statistics/.

The internet's undersea world. 2008. *The Guardian,* February 2.

iPass Inc. 2007. *Ipass Network Diagram*. http://www.ipass.com/technology/index
.html.

ITU. 2006. *ITU Internet Reports 2005: The Internet of Things*. http://www.itu.int/osg/
spu/publications/internetofthings/.

I Want One Of Those Ltd. 2008. *WiFi Watch—I Want One of Those*. http://www
.iwantoneofthose.com/wifi-watch/index.html.

James, William. 1909. *The Meaning of Truth: A Sequel to "Pragmatism"*. New York:
Longmans, Green.

James, William. 1911. *Some Problems of Philosophy: A Beginning of an Introduction to
Philosophy*. New York: Longmans, Green.

James, William. 1960. *The Will to Believe, and Other Essays in Popular Philosophy, and
Human Immortality*. New York: Dover.

James, William. 1996a. *Essays in Radical Empiricism*. Lincoln: University of Nebraska
Press.

James, William. 1996b. *A Pluralistic Universe: Hibbert Lectures at Manchester College on
the Present Situation in Philosophy*. Lincoln: University of Nebraska Press. (Originally
published by Longmans, Green, New York, 1909.)

James, William. 2004. *Pragmatism: A New Name for Some Old Ways of Thinking*. http://
www.gutenberg.org/dirs/etext04/prgmt10.txt.

James, William. 1975. *The Meaning of Truth: The Works of William James*. Ed. Fredson
Bowers and Ignas K. Skrupskelis. Cambridge, MA: Harvard University Press.

James, William. 1978. *Pragmatism: A New Name for Some Old Ways of Thinking;
The Meaning of Truth: A Sequel to Pragmatism*. Ed. Fredson Bowers and Ignas K.
Skrupskelis. Cambridge, MA: Harvard University Press.

Jhai Foundation. 2003. *The Remote IT Village Project*. http://www.jhai.org/jhai_
remoteIT.html.

JiWire Inc. 2008a. *JiWire Wi-Fi Advertising Formats for Service Providers*. http://www
.jiwire.com/providers/ad-formats.htm.

JiWire Inc. 2008b. *JiWire Wi-Fi Advertising Network Locations*. http://www.jiwire.com/
advertisers/ad-network-locations.htm.

JiWire Inc. 2008c. *JiWire Wi-Fi Finder and Hotspot Directory*. http://www.jiwire.com/.

Joel, Micah. 2004. *Safe and Insecure. I Opened Up My Wireless Home Network to the
World, and I've Never Felt More Comfortable*. May 18. http://www.salon.com.

Jones, Neil C., and Pavel Pevzner. 2004. *An Introduction to Bioinformatics Algorithms:
Computational Molecular Biology*. Cambridge, MA: MIT Press.

Kauffman, Stuart A. 1995. *At Home in the Universe: The Search for Laws of Self-Organization and Complexity*. New York: Oxford University Press.

Kershaw, Mike. 2008. *Kismet*. http://www.kismetwireless.net/.

Kewney, Guy. 2004. *e-Minister Will Make Every Public Library a Wi-Fi Hotspot*. http://www.theregister.co.uk/content/69/34352.html.

Kewney, Guy. 2004. *Wireless Lamp Posts Take Over World*. January 15. http://www.theregister.co.uk/content/69/34894.html.

Kim, Ryan. 2007. *World's Largest Wi-Fi Having Growing Pains: Taipei's System Began 15 Months Ago, But Subscriptions Have Fallen Far Short of What City, Contractor Had Expected*. Hearst Communications, April 9. http://www.sfgate.com/cgi-bin/article.cgi?file=/chronicle/archive/2007/04/09/BUGK8P4DCO1.DTL&type=printable.

Kittler, Friedrich. 1993. Real time analysis, time axis manipulation. In *Draculas Vermchtnis Technische Schriften*. Leipzig: Reclam Verlag.

Knorr-Cetina, Karin, and Urs Bruegger. 2002. Traders' engagement with Markets: A postsocial relationship. *Theory, Culture & Society* 19 (5–6):161–185.

Kostof, Spiro. 1992. *The City Assembled: The Elements of Urban Form through History*. Boston: Little, Brown.

Kotadia, Munir. 2005. *First Wireless Feast Hits UK Market*. http://www.silicon.com/networks/mobile/0,39024665,39119356,00.htm.

Kwastek, Katja, Inke Arns, Nina-Maria Faulstich, and Cuxhavener Kunstverein. 2004. *Ohne Schnur: Kunst und drahtlose Kommunikation = Art and Wireless Communication*. Cuxhaven: Cuxhavener Kunstverein.

Langton, Charan. 2007. *Tutorial 12: Coding and Decoding with Convolutional Codes*. http://complextoreal.com/chapters/convo.pdf.

Lapoujade, David. 2000. From transcendental empiricism to worker nomadism: William James. *Pli* 9:190–199.

Latour, Bruno. 1999. *Pandora's Hope: Essays on the Reality of Science Studies*. Cambridge, MA: Harvard University Press.

Latour, Bruno. 2007. La Connaissance est-elle un mode d'existence? Recontre au muséum de James, Fleck et Whitehead avec des fossiles de chevaux. In *Vie et Expérimentation Peirce, James, Dewey*, ed. D. Debaise. Paris: Libraire Philosophique J.Vrin.

Laurier, Eric. 2004. *The Cappuccino Community: Cafes and Civic Life in the Contemporary City*. ESCR-funded project, University of Glasgow, https://dspace.gla.ac.uk/retrieve/142/elaurier002.pdf.

Lazzarato, Maurizio. 1998. *New forms of production and circulation of knowledge*. Nettime.http://www.nettime.org/Lists-Archives/nettime-l-9810/msg00113.html

Lazzarato, Maurizio. 2002. *Puissances de l'invention: La psychologie âeconomique de Gabriel Tarde contre l'âeconomie politique.* Paris: Empêcheurs de penser en rond.

Lesk, Arthur M. 2002. *Introduction to Bioinformatics.* Oxford: Oxford University Press.

Leyden, John. 2004. *Wireless Access Point Triggers Bomb Scare.* October 29. http://www.theregister.co.uk/2004/10/29/wireless_ap_bomb_scare/.

Leyden, John. 2007. *Drive-by Wi-Fi "Thief" Heavily Fined.* Channel Register. http://www.channelregister.co.uk/2007/05/23/michigan_wifi_conviction/.

Linksys WRT54G series. 2008. *Wikipedia.* http://en.wikipedia.org/wiki/Wrt54g.

LocustWorld. 2004. *The Information Revolution: Mesh Networking Hardware and Software.* http://www.locustworld.com/.

Daily Mail Online. 2008. *London to Get Free Wi-Fi between Central London and Greenwich.* http://www.dailymail.co.uk/sciencetech/article-468758/London-free-Wi-Fi -central-London-Greenwich.html.

Collective, London Hacklabs. 2008. *Media Hacklab.* http://www.hacklab.org.uk/.

Lovink, Geert. 2003. Hi-low: The bandwidth dilemma, or Internet stagnation after dotcom mania. In *Dark Fiber: Tracking Critical Internet Culture.* Cambridge, MA: MIT Press.

Macdonald, Nico. 2004. *Wi-Fi in the Real World.* Pt. 1. February 11. http://www .theregister.co.uk/content/69/35461.html.

Mackenzie, Adrian. 2002. *Transductions: Bodies and Machines at Speed. Technologies.* London: Continuum.

Mackenzie, Adrian. 2005. Untangling the unwired: Wi-Fi and the cultural inversion of infrastructure. *Space and Culture* 8 (3):269–285.

Mackenzie, Adrian. 2006a. Cutting code: software and sociality. In *Digital Formations,* ed. S. Jones. New York: Peter Lang.

Mackenzie, Adrian. 2006b. Innumerable transmissions: Wi-Fi® from spectacle to movement. *Information Communication and Society* 9 (6):779–800.

Mackenzie, Adrian. 2006c. The meshing of impersonal and personal forces in technological action. *Culture, Theory & Critique* 47 (2):197–212.

Mackenzie, Adrian. 2008. Codecs. In *Software Studies,* ed. M. Fuller. Cambridge, MA: MIT Press.

Mackenzie, Adrian. 2009. Codecs: From centres of calculation to centres of envelopment. In *Deleuze in Science and Technology Studies,* ed. C. B. Jensen and K. Rodje. Oxford: Berghahn Publishers.

Make: Technology on Your Time. 2008. O'Reilly Media Inc. http://makezine.com/magazine/.

Mannion, Patrick, and Mike Clendenin. 2004. *China's Wi-Fi Security Stance Ruffling Feathers.* http://www.commsdesign.com/printableArticle/;jsessionid=J4TY4V2NY5G FKQSNDBCCKHQ?articleID=17000455.

Marriot, Michael. 2006. Hey neighbor, stop piggybacking on my wireless. *New York Times*, March 6, http://www.nytimes.com/2006/03/05/technology/05wireless.html?_r=1&oref=slogin.

Massumi, Brian. 2000. Too-blue: Colour-patch for an expanded empiricism. *Cultural Studies* 14 (2):177–226.

Massumi, Brian. 2002. *Parables for the Virtual.* Durham, N.C: Duke University Press.

Mattelart, Armand. 1996. *The Invention of Communication.* Minneapolis: University of Minnesota Press.

Mattelart, Armand. 2000. *Networking the World, 1794–2000.* Minneapolis: University of Minnesota Press.

Maxwell, James Clerk. 1865. A dynamical theory of the electromagnetic field. *Philosophical Transactions of the Royal Society of London* 155:459–512.

McCarthy, Kieren. 2004. *Have Fun with Wi-Fi in a Rucksack.* March 24. http://www.theregister.co.uk/2004/03/24/have_fun_with_wifi/.

McIntosh, Neil. 2003. Will Wi-Fi fly? *The Guardian*, August 13, http://www.guardian.co.uk/print/0,3858,4732128-110837,00.html.

Networks, Meraki. 2008a. *Free the Net.* http://sf.meraki.com/.

Networks, Meraki. 2008b. *Meraki Wireless Network: Affordable Internet Solution: Free Wi-Fi.* http://meraki.com/.

Merrit, Rick. 2008. *Video: Berkeley Brings WiFi to Rural Poor.* June 17. http://www.eetimes.com/news/latest/showArticle.jhtml;jsessionid=Y2QN4JAPNQB02QSNDLPS KH0CJUNN2JVN?articleID=208700017.

Miller, Daniel, and Don Slater. 2000. *The Internet: An Ethnographic Approach.* Oxford: Berg.

Miller, Stewart S. 2003. *Wi-Fi Security.* New York: McGraw-Hill.

Mills, Elinor. 2007. *Google in San Francisco: "Wireless Overlord"?* http://www.news.com/2100-1039_3-5886968.html.

Mills Abreu, Elinor. 2004. *Coffee, Tea or WiFi? PluggedIn: Web Access in the Clouds.* http://www.reuters.com/.

Mims, Christopher. 2008. *Meraki's Guerilla Wi-Fi to Put a Billion More People Online.* http://www.sciam.com/article.cfm?id=merakis-guerilla-wi-fi-to-put-billion-people-online.

Mitchell, William J. 2003. *Me++: The Cyborg Self and the Networked City.* Cambridge, MA: MIT Press.

Mo, Yan-chih. 2005. *TAIWAN: Ma Says Taipei to Be Wireless in Two Days.* http://www.asiamedia.ucla.edu/article.asp?parentid=35839.

Mullarkey, John. 1999. *Bergson and Philosophy.* Edinburgh: Edinburgh University Press.

Muller, Nathan J. 2003. *WiFi for the Enterprise.* New York: McGraw-Hill.

Munster, Anna, and Geert Lovink. 2005. Theses on distributed aesthetics: Or, what a network is not. *FibreCulture* 7. http://journal.fibreculture.org/issue7/issue7_munster_lovink.html.

MyZones. 2003. *Lose the Wires, Be Free. Promotional brochure.* Manchester: MyZones Ltd.

Low-cost telecoms: Yabba dabba do.2008. *The Economist,* July 17.

Nancy, Jean-Luc. 2000. *Being Singular Plural.* Stanford, CA: Stanford University Press.

Nancy, Jean-Luc. 2007. *The Creation of the World, or, Globalization.* Albany: State University of New York Press.

Negroponte, Nicholas. 2002. *Being Wireless* (10.1). http://www.wired.com/wired/archive/10.10/wireless.html.

NetHope LLC. 2007. *NetHope: Wiring the Global Village.* http://www.nethope.org/index.html.

Newman, M. E. J. Albert-László Barabási, and Duncan J. Watts. 2006. *The Structure and Dynamics of Networks.* Princeton Studies in Complexity. Princeton, NJ: Princeton University Press.

NoCat. 2008. *NoCatNet.* http://nocat.net/.

Node D. B. com. 2008. *About NodeDB.com.* http://www.nodedb.com/about.php?id=0.

NYCWireless. 2008. *Using the Public Airwaves to Connect and Strengthen Communities in New York City.* http://www.nycwireless.net/about/.

O3B Networks. 2008. *Connecting the Other 3 Billion* http://www.o3bnetworks.com/index.html.

OLPC. 2008. *Social Sharing.* http://www.laptopgiving.org/en/social-sharing.php#slide02.

Open Handset Alliance. 2009. *What Would It Take to Build a Better Mobile Phone?* http://www.openhandsetalliance.com/.

OpenPark.net. 2004. *The Open Park Project (Open Park): WiFi Internet Access on the National Mall.* http://www.openpark.net/.

openspectrum.info. 2007. *OpenSpectrum.Info: Africa.* http://www.volweb.cz/horvitz/os-info/africa.html.

OpenWRT. 2008. *OpenWRT Wireless Freedom.* http://openwrt.org/.

Ouestaf News. Télécoms béninois : anarchie totale, mesures draconiennes annoncées. 2007. *Ouestaf.com*, January 20. http://www.ouestaf.com/Telecoms-beninois-anarchie-totale,-mesures-draconiennes-annoncees_a267.html.

Panesar, G., D. Towner, A. Duller, A. Gray, and W. Robbins. 2005. Deterministic parallel processing. *International Journal of Parallel Programming* 34:323–341.

Patrick, Ugeh. 2008. *Interweb Satcom, Alvarion to Deploy WIMAX Techa.* http://www.thisdayonline.com/nview.php?id=103821.

Paul, John. 2006. *Google's Big BOP Bet? Bringing Wi-Fi to Africa.* http://www.icicemac.com/nouvelle/index.php3?nid=7049.

PIC. 2005. *HOWTO: Make That PalmOne Treo 650 Even Better!* http://www.palminfocenter.com/print.asp?ID=7443&s=1.

PicoChip. 2007. *PC202 Integrated Baseband Processor Product Brief.* http://www.picochip.com/downloads/03989ce88cdbebf5165e2f095a1cb1c8/PC202_product_brief.pdf.

Pignarre, Philippe, and Isabelle Stengers. 2005. *La sorcellerie capitaliste: Pratiques de désenvoûtement.* Paris: Découverte.

Pogue, David. 2009. *State of the Art—With a Private MiFi Hot Spot, Be Online Wherever You Like.* NYTimes.com. http://www.nytimes.com/2009/05/07/technology/personaltech/07pogue.html.

Poulsen, Kevin. 2004. *Wireless Hacking Bust in Michigan.* http://www.theregister.co.uk/content/69/33959.html.

Povoledo, Elisabetta. 2006. Wireless: The tin-can antenna offers a boon for third world. *International Herald Tribune*, March 1.

PPA. 2004. *The Pico Peering Agreement v. 1.0.* http://picopeer.net/wiki/index.php/PicoPeeringAgreement.

Priest, Julian. 2005. *The State of Wireless London.* http://informal.org.uk/people/julian/publications/the_state_of_wireless_london/.

Puzzangherra, Jim. 2008. *FCC Votes to Turn Empty TV Channels into Wireless Net Access*. November 4. http://latimesblogs.latimes.com/technology/2008/11/federal-regulat.html.

Qualcomm. 2005. *About Qualcomm*. http://www.qualcomm.com/.

Qware Communications. 2008. *WiflyNet* http://www.wifly.com.tw/wifly6/tw/WiflyNet/AboutWiflyNet/WiflyNetAdvantage/.

Rabinow, Paul. 2003. *Anthropos Today: Reflections on Modern Equipment*. Princeton, NJ: Princeton University Press.

Rajan, Kaushik Sunder. 2003. Genomic capital: Public cultures and market logics of corporate biotechnology. *Science as Culture* 12 (1):87–121.

Rajan, Kaushik Sunder. 2006. *Biocapital: The Constitution of Postgenomic Life*. Durham, NC: Duke University Press.

Rasch, Mark. 2004. *The Wi-Fi User as Wireless Felon*. May 4. http://www.theregister.co.uk/2004/05/04/us_wi_fi_legislation/.

Raymond, Eric. 2003. *brick*. The Jargon File 4.4.7. http://www.catb.org/jargon/html/B/brick.html.

Raymond, Eric. 1996. *The New Hacker's Dictionary*. Cambridge, MA: MIT Press.

Redding, Paul. 2001. Embodiment, conceptuality and intersubjectivity. *Journal of Speculative Philosophy* 15 (4):257–271.

Rethink Associates. 2005. *Wireless Watch: In Depth Analysis of Wlan, Cellular and Broadband Wireless Markets*. London: Rethink Associates.

Rheingold, Howard. 2002. *Smart Mobs: The Next Social Revolution*. Cambridge, MA: Perseus.

Robbins, Alexandra. 2004. *Nosy Wi-Fi Neighbors*. http://www.pcmag.com/print_article/0,1761,a=112257,00.asp.

Rodowick, David Norman. 2001. *Reading the Figural, or, Philosophy After the New Media*. Durham, NC: Duke University Press.

Ross, Alison. 2007. *The Aesthetic Paths of Philosophy: Presentation in Kant, Heidegger, Lacoue-Labarthe, and Nancy*. Stanford, CA: Stanford University Press.

Ross, Andrew. 1991. *Strange Weather: Culture, Science and Technology in the Age of Limits*. London: Verso Books.

Rwanda News Agency/Agence Rwandaise d'Information. 2007. *Rwanda: UN Chief Commends Kigali "Connect Africa" Plan*. July 13. http://allafrica.com/stories/200707131041.html.

Ryman, Geoff. 2004. *Air, or, Have Not Have*. New York: St. Martin's Griffin.

Savicic, Gordan. 2008. *Constraint City*. http://www.yugo.at/equilibre/.

SeattleWireless. 2004. *Seattle Wireless TV*. http://tv.seattlewireless.net/.

Serres, Michel, and Lawrence R. Schehr. 1982. *The Parasite*. Baltimore: Johns Hopkins University Press.

Shapiro, Lynn. 2009. *DOTmed.com—St. Jude's Wi-Fi Pacemaker Wins Approval*. http://www.dotmed.com/news/story/9878/.

Shaviro, Steven. 2003. *Connected, or, What It Means to Live in the Network Society*. *Electronic Mediations*. vol. 9. Minneapolis: University of Minnesota Press.

Simondon, Gilbert. 1989. *Du mode d'existence des objets techniques*. Paris: Aubier.

Simone, Abdou Maliq. 2006. Intersecting Geographies? ICTs and Other Virtualities in Urban Africa. In G. F. Downey and Melissa S. Fisher, eds., *Frontiers of Capital: Ethnographic Reflections on the New Economy*. Durham: Duke University Press.

SkyPilot Networks. 2009. *Automatic Meter Reading*. http://www.skypilot.com/pdf/app_autometerreading.pdf.

Slater, Don, and Janet Kwami. 2005. *Embeddedness and Escape: Internet and Mobile Use as Poverty Reduction Strategies in Ghana*. Information Society Research Group (ISRG) http://zunia.org/uploads/media/knowledge/internet.pdf.

Sloterdijk, Peter. 2004. *Plurale Spärologie, 1: Sphären*. Frankfurt am Main: Suhrkamp.

Smith, Steven W. 2004. *The Scientist and Engineer's Guide to Digital Signal Processing*. San Diego, CA: California Technical Publishing.

Smith, Tony. 2003. *London Gets UK's First Wi-Fi "Hotzone."* The Register, http://www.theregister.co.uk/2003/11/10/london_gets_uks_first_wifi/.

Smith, Tony. 2005a. *China Agrees to Drop WAPI Wireless Spec*. The Register. http://www.theregister.co.uk/2004/04/22/china_caves_on_wapi/.

Smith, Tony. 2005b. *WiMAX Hype Peaks*. The Register. http://www.theregister.co.uk/2005/03/24/wimax_gartner_outlook/.

Snellgrove, Andrew, and Tim Hearn. 2006. *The Wireless City: A Showcase* http://www.westminster.gov.uk/councilgovernmentanddemocracy/councils/modernisation/westminsterwirelesscity/upload/36080_1.pdf.

Standage, Tom. 2003. Beyond the telecoms bubble: A survey. *Economist* 11 (October):13–24.

St. Clair, Richard. 2004. *Creating a Wireless Nation*. http://www.niue.nu/images/Nuiepaper38.pdf.

Stengers, Isabelle. 2000. *The Invention of Modern Science. Theory Out of Bounds*. vol. 19. Minneapolis: University of Minnesota Press.

Stengers, Isabelle. 2005. Deleuze and Guattari's last enigmatic message. *Angelaki* 10 (1):151–167.

Strathern, Marilyn. 2006. Imagined collectivities and multiple authorship. In *Code: Collaborative Ownership and the Digital Economy*, ed. Rishab Aiyer Ghosh. Cambridge, MA: MIT Press.

Swan, Stan. 2005. *USB Adaptors & DIY Antenna = "Poor Man's WiFi"?* http://www .usbwifi.orcon.net.nz/.

Sykes, K. 2004. *The Malanggan's Claim: Ethics, Aesthetics and Property Relations in New Ireland Public Art*. http://www.socialsciences.manchester.ac.uk/disciplines/ socialanthropology/research/workingpapers/documents/Malanggan%20as%20 Public%20Art%20rev.pdf.

Taipei: World's Largest Wifi Grid. 2004. http://story.news.yahoo.com/news?tmpl=sto ry&cid=575&ncid=738&e=6&u=/nm/20041119/wr_nm/tech_taiwan_cybercity_dc.

Tarde, Gabriel. 1902. *Psychologie économique*. Paris: F. Alcan.

Tarkka, Minna. 2008. *Labours of Location: Acting in the Pervasive Media Space*. http:// diffusion.org.uk/?p=105.

Frontières, Télécoms Sans. 2007. *Communications for Life*. http://www.tsfi.org/ tsfispip/rubrique.php3?id_rubrique=37&lang=en.

Telegeography Research. 2007. *Global Bandwidth Executive Summary*. http:// telegeography.com/products/gb/pdf/Executive_Summary.pdf.

Terranova, Tiziana. 2000. Free labor: Producing culture for the digital economy. *Social Text* 63 (18):33–57.

Terranova, Tiziana. 2004. *Network Culture: Politics for the Information Age*. London: Pluto Press.

Thrift, Nigel. 2004a. Movement-space: The changing domain of thinking resulting from the development of new kinds of spatial awareness. *Economy and Society* 33 (4):582–604.

Thrift, Nigel. 2004b. Remembering the technological unconscious by foregrounding knowledges of position. *Environment and Planning. D, Society & Space* 22 (1):175–191.

TIER. 2008. *Technology and Infrastructure for Emerging Regions*. http://tier.cs.berkeley. edu/wiki/Home.

Tonkiss, Fran. 2005. *Space, the City and Social Theory: Social Relations and Urban Forms*. Cambridge, MA: Polity Press.

Toshiba Corporation. 2003. *Enter the World of Freedom Computing. Promotional brochure*. New York, N.Y.: Toshiba Corporation.

Townsend, Anthony. 2000. Life in the realtime city: Mobile telephones and urban metabolism. *Journal of Urban Technology* 7 (2):85–104.

Trillo, Richard. 2008. *Mali: In Touch with Timbuktu.* http://www.roughguides.com/website/Travel/SpotLight/ViewSpotLight.aspx?spotLightID=399.

Tropos Networks. 2006a. *Applications Overview.* Tropos Networks, #196. http://www.tropos.com

Tropos Networks. 2006b. *The Proven Leader in Delivering Ubiquitous, Metro-Scale Wi-Fi Mesh Network Systems.* http://www.tropos.com/.

Tsing, Anna Lowenhaupt. 2005. *Friction: An Ethnography of Global Connection.* Princeton, NJ: Princeton University Press.

Tyler Hamilton, N. 2006. *T.O. to Become Wireless Hotspot.* March 6. http://www.thestar.com/NASApp/cs/ContentServer?pagename=thestar/Layout/Article_Type1&c=Article&cid=1141643034143&call_pageid=968332188492&col=968793972154&t=TS_Home.

United Nations. 2004. *Conference on Wireless Internet Opportunity for Developing Nations at UN Headquarters 26 June 2003. Press Release Note No. 5799.* http://www.un.org/News/Press/docs/2003/note5799.DOC.htm.

United Villages Inc. 2008. *Access for All.* http://www.unitedvillages.com/.

Van Dijk, Jan. 2006. *The Network Society.* 2nd ed. Thousand Oaks, CA: Sage.

Vietnam News Agency. 2007. *WiMax Lays Plans for Broader City Coverage.* http://vietnamnews.vnagency.com.vn/showarticle.php?num=02BUS151007.

Virno, Paolo. 2004. *A Grammar of the Multitude: For an Analysis of Contemporary Forms of Life.* Semiotext(e) Foreign Agents Series. Los Angeles: Semiotext(e).

Viterbi algorithm. 2008. *Wikipedia.* http://en.wikipedia.org/wiki/Viterbi_algorithm.

Viterbi, Andrew J. 1967. Error bounds for convolutional codes and an asymptotically optimum decoding algorithm. *IEEE Transactions on Information Theory* 13 (2):260–267.

Vladimirov, Andrew A., Konstantin V. Gavrilenko, and Andrei A. Mikhailovsky. 2004. *Wi-Foo.* Boston: Addison-Wesley.

Wainwright, Martin. 2003. The future is nearly in sight. *The Guardian,* July 31, 21–22.

Wakeford, Nina. 2003. The embedding of local culture in global communication: Independent Internet cafés in London. *New Media & Society* 5 (3):379–399.

Waltner, Charles. 2004. *Cisco Donates Equipment to Build the World's Highest Wireless Connection on Mount Everest.* http://newsroom.cisco.com/dlls/hd_012303.html.

Wark, McKenzie. 2004. *A Hacker Manifesto*. Cambridge, MA: Harvard University Press.

WCM. 2003. *The Wireless Commons Manifesto*. http://www.wirelesscommons.org/history/manifesto.html.

Wearden, Graeme. 2008. *London Gets Monster Wi-Fi Hot Spot*. http://news.zdnet.co.uk/communications/0,1000000085,39117757,00.htm.

Weber, Tim. 2003. *Wi-Fi Will Be "Next Dot.com Crash."* BBC News UK Edition. June 20. http://news.bbc.co.uk/1/hi/business/3006740.stm.

WeFi.com. 2008a. *WeFi—Find Free Wi-Fi Anywhere and Friends Everywhere!* http://www.wefi.com/.

WeFi.com. 2008b. *Wefi: Making WiFi Easy*. http://www.wefi.com/wicket/main.html.

Werbach, Kevin. 2003. *Radio Revolution: The Coming Age of Unlicensed Wireless*. http://werbach.com/docs/RadioRevolution.pdf.

Westminster City Council. 2006a. *Westminster Goes WiFi with BT*. http://www.westminster.gov.uk/councilgovernmentanddemocracy/councils/modernisation/westminsterwirelesscity/index.cfm.

Westminster City Council. 2006b. *Westminster Pilots "the Wireless City."* http://www.westminster.gov.uk/news/PR-1853.cfm.

Whisher Solutions. 2008. *Whisher: Building the World's Largest WiFi Network*. http://www.whisher.com/.

Wi-Fi Alliance. 2004. *Zone Finder*. http://www.wi-fizone.org.

Wi-Fi Alliance. 2008. *Wi-Fi Alliance: Get to Know the Alliance*. http://www.weca.net/about_overview.php.

Wi-Fi Alliance. 2009. *Wi-Fi Alliance—Our Brand*. http://www.wi-fi.org/brand_usage.php.

WiFiinschools. 2009. *Home—WiFiinschools.org.uk*. http://wifiinschools.org.uk/index.html.

Wi-Fi TV. 2008. *Wi-Fi TV Social Internet TV*. http://www.wi-fitv.com/index.php.

Wild Packets Solutions. 2005. *Airopeek NX*. http://www.wildpackets.com/solutions/wireless.

Williams, Martyn. 2004. *Aibo Gets a Face-Lift*. http://www.pcworld.com/resource/printable/article/0,aid,112324,00.asp.

wimaxday. 2008. *OTE Launches WiMAX on the "Holy Mountain": WiMAX Day*. http://www.wimaxday.net/site/2008/08/20/ote-launches-wimax-on-the-holy-mountain/.

wire.less.dk. 2008. *Wireless Roadshow*. March 23. http://wire.less.dk/wiki/index.php/Main_Page#Wireless_Roadshow.

Wireless London. 2006. *Welcome to Wireless London*. http://wirelesslondon.info/HomePage.

Wireless Philadelphia Executive Committee. 2006. *A 21st Century Opportunity*. Philadelphia: Wireless Philadelphia Executive Committee.

WLAN. 2003. *WLAN Event, Olympia Exhibition Centre, London, May 23–24, 2003*. http://www.wlanevent.com.

Wolfowitz, Paul. 2004. *Directive Number 8100.2 Use of Commercial Wireless Devices, Services, and Technologies in the Department of Defense (DoD) Global Information Grid (GIG)*. April 14. http://www.dtic.mil/dticasd/sbir/sbir041/srch/n076.pdf.

Zachary, G. Pascal. 2008. *Inside Nairobi, the Next Palo Alto?* http://www.nytimes.com/2008/07/20/business/worldbusiness/20ping.html?ex=1217217600&en=18779b83fccc23b0&ei=5070&emc=eta1.

Zuniga, Ricardo Miranda. 2004. *The Public Broadcast Cart*. http://www.ambriente.com/wifi/index.html.

Index